Epistemological Approaches to Digital Learning in Educational Contexts

Epistemological Approaches to Digital Learning in Educational Contexts is dedicated to topical issues in school education and pedagogical science related to the learning process in a technology- and media-enriched environment. It opens up discussions on the development of the educational science sector and strategies for smart pedagogy to promote synergy between technology and pedagogy to support students in the learning process.

The book presents different perspectives on how to evaluate the enhancement of technology use, which can help improve computational thinking skills. It also helps in identifying the changes in pupils' algorithmic thinking through programming in Scratch 2.0. The book further explores the way digitally mediated materiality may support teaching practice and proposes tools that are available for the educational curator in a digital learning environment.

This book will be of great interest to academics, researchers, and post-graduate students in the fields of higher education, vocational education, and digital learning.

Linda Daniela is professor, Chair of the PhD Defense Board in Educational Sciences, Head of the Scientific Institute of Pedagogy at the University of Latvia in Riga, Latvia.

Perspectives on Education in the Digital Age

Series Editors: David Kergel and Birte Heidkamp

The process of digitalization is leading to a fundamental social change affecting all spheres of social life. In the pedagogical field there is a need for re-structuring key concepts such as learning, teaching and education that considers socio-economic and cultural changes.

Perspectives on Education in the Digital Age explores the process of coming to terms with socio-economic and socio-cultural shifts arising from digitalization and discusses this process with reference to its effects on education. The Series provides a forum for discussion of critical, integrative analyses of social transformations in the digital age, drawn from different fields such as the humanities, social sciences and economics. The aim of the Series is to analyse the implications of cultural change on education in the digital age by bringing together interdisciplinary dialogue and different theoretical approaches.

The Hidden Curriculum of Online Learning
Understanding Social Justice through Critical Pedagogy
Murat Oztok

Communication and Learning in an Age of Digital Transformation
Edited by David Kergel, Birte Heidkamp, Ronald Arnett, and Susan Mancino

Pedagogies of Digital Learning in Higher Education
Edited by Linda Daniela

Epistemological Approaches to Digital Learning in Educational Contexts
Edited by Linda Daniela

For more information about this series, please visit: https://www.routledge.com

Epistemological Approaches to Digital Learning in Educational Contexts

Edited by
Linda Daniela

Routledge
Taylor & Francis Group
LONDON AND NEW YORK

First published 2020
by Routledge
2 Park Square, Milton Park, Abingdon, Oxon OX14 4RN

and by Routledge
52 Vanderbilt Avenue, New York, NY 10017

Routledge is an imprint of the Taylor & Francis Group, an informa business

British Library Cataloguing-in-Publication Data
A catalogue record for this book is available from the British Library

Library of Congress Cataloging-in-Publication Data
A catalog record has been requested for this book

ISBN: 9780367333799 (hbk)
ISBN: 9780429319501 (ebk)

Typeset in Bembo
by codeMantra

Contents

Figures

Figures

Tables

Contributors

Samuele Calzone's areas of expertise include pedagogy and philosophy of education. In particular, he is focused on the introduction of ICT at school, with regard to teacher digital skills and media education and school governance. He is working on tailored learning in adult education, as well as leading a project entitled "monitoring and research" on the results of the National Operational Program (NOP) on education.

Miroslava Černochová is a teacher educator at the Faculty of Education, Charles University. She is a guarantor of the MA programme "*Teaching of subjects for primary and secondary education – Technology and Informatics Education*". She is also a teacher of the course, "Didactics of information technology in primary education". She has participated in a number of EU projects (2AgePro, DIYLab, Images and Identity: Improving Citizenship through Digital Media, etc.). She is a member of IFIP WG "Informatics in Education" and a chair of the ATEE RDC "Teacher Education and Digital Technology". She was delegated to participate in some UNESCO working meetings dedicated to digital technology or computational thinking in (teacher) education. Since 2017 she is an institutional manager of the national project PRIM (imysleni.cz) focused on computational thinking development of students and teachers in schools and student teachers, too.

Linda Daniela is professor, Chair of the PhD Defense Board in Educational Sciences, Head of the Scientific Institute of Pedagogy at the University of Latvia. Her expertise spans technology-enhanced learning, smart pedagogy, smart education, educational technologies for learning, educational robotics, etc. She is an author and co-author of more than 70 publications about processes in all dimensions of education. She has been involved in more than 30 research projects, editing books and journals on technological aspects in education.

Ioannis Deliyannis holds a BSc computer science and PhD in the field of interactive multimedia (2002) at the computer science department of

University of Wales at Swansea. He is assistant professor of interactive and multimedia communication technologies and faculty member of the Department of Audiovisual Arts at Ionian University, Corfu, Greece.

Emmanuel Fokides is an assistant professor in the Department of Primary School Education, University of the Aegean, Greece. His courses focus on the educational uses of emerging technologies, virtual reality, digital storytelling, augmented reality, and serious games. Since 1994, he is involved in a number of research projects regarding distance and lifelong learning and the educational uses of virtual and augmented reality. His work is published in several conference proceedings, chapters in edited books, and journals.

Marilene Santana dos Santos Garcia holds a bachelor's degree in pedagogy (UNINTER), in German, Portuguese and Linguistics (USP), master's degree (UNICAMP), doctor degree (USP/University of Oldenburg – Germany), post-doctor studies (TIDD – PUC-SP). She is a professor at Master's Degree in Education and New Technologies – UNINTER. She has authored the following books: "Technological Mobility and Didactic Planning" (2017), "Evaluation and Validation of Projects" (2018), "Educational Curatorship" (2019)

Polyxeni Kaimara is a developmental and educational psychologist with an MSc degree in public health, and is pursuing a PhD at the Department of Audiovisual Arts of the Ionian University, Greece. She is also a *certified* counselor and adult trainer. Her interests focus on the design of digital materials for special and inclusive education.

Giuseppina Rita Jose Mangione is a researcher (tenured position) at INDIRE (National Institute of Documentation, Innovation, and Educational Research). She holds a master's degree in communication sciences and a PhD in telematics and information society obtained at the University of Florence. She has been a postdoctoral associate at the University of Salerno, working on e-learning and innovative educational methods. Her research interests in INDIRE include teachers' training in a *small school*, multigrade teaching methods and technologies for innovation aimed at overcoming cultural isolation. She managed several projects in the field of education and small schools, collaborating with international institutions and governmental agencies. She is author of several books and about one-hundred scientific papers. Personal page: www.indire.it/personale/giuseppina-rita-mangione/

Luciano Frontino de Medeiros earned his PhD in knowledge engineering and management at UFSC, Brazil. He is currently professor of applied artificial intelligence at the Post-graduation Program in Education and New Technologies, UNINTER, Brazil, and head of the computing simulations and educational robotics research group.

Divya Menon, currently pursuing a MSc CreaSmartEdTech from Université Côte d'Azur, is an independent education consultant who works with various international organizations to help design and implement impactful learning experiences. She's doing an internship on the role of artificial intelligence devoted to education with Inria Labs and LINE laboratory.

Sofia-Maria Poulimenou received a BA in international and European relations and an MSc in international law and diplomacy in 2006, from the Department of International and European Studies, Panteion University. She is a PhD candidate in World Heritage Sites Management at the Department of Archives, Library Science and Museology, Ionian University, Corfu, Greece.

Margarida Romero is research director of the Laboratoire d'Innovation et Numérique pour l'Éducation (LINE), a research lab in the field of technology-enhanced learning (TEL), full professor at Université Côte d'Azur (France), and associate professor at Université Laval (Canada). Her research is oriented towards the inclusive, humanistic, and creative uses of technologies (co-design, game design, and robotics) for the development of creativity, problem solving, collaboration, and computational thinking.

Hasan Selcuk, PhD, is a postdoctoral researcher and lecturer at the Faculty of Education, Charles University in Prague. He earned his PhD in education from King's College London in 2017. Among his topics of research interest are self-assessment of ICT and foreign-language skills of secondary school students, digital creativity, and autonomous learning.

Sowmya BP is currently pursuing a MSc CreaSmartEdTech from Université Côte d'Azur. She is working towards initiating the Class'Code project in India. She designs and develops learning solutions for BCD Travel. She is an external collaborator for the project on the role of artificial intelligence devoted to education with Inria Labs and LINE laboratory.

Milan Svoboda is an ICT and art teacher at a primary school in Prague. He graduated from the Faculty of Education of Charles University, where he studied Information and Communication Technology.

Thierry Viéville is a senior researcher at Inria (National Research Institute in Computer Science and Control Theory) where he works in Computational Neurosciences, in the mnemosyne research team and in the LINE laboratory regarding education in computer science.

Jennifer Wong-Powell is passionate about unearthing the complexities of play in the digital age. She believes that through collaboration and conversation at a global stage we can begin to develop a shared way of thinking, communicating, and supporting digital play in early childhood education.

Preface

Epistemological approaches to digital learning in educational contexts

This book is dedicated to topical issues in school education and pedagogical science related to the learning process in a technology- and digital media-enriched environment, opening up discussions on the development of the educational science sector and strategies for smart pedagogy to promote synergy between technology and pedagogy to support students in the learning process.

There are two important milestones in digital transformation that also influenced education: The year 1971, which can be considered as the beginning of the digital revolution, when Intel introduced a microprocessor that enabled the rapid development of computer technology, which gradually entered the educational environment; and the year 1993, when the World Wide Web (WWW) became freely available and when the era of free information flow started. At the beginning, there were assumptions that freely available information would ensure that students would learn in a self-regulatory manner, obtaining the necessary information for their knowledge construction anywhere and anytime. More than two decades have passed since then, and it has become clear that neither digital technology nor free access to information can facilitate a rapid leap in the growth of knowledge. Gradually, since the phase of overall fascination, hoping for technological progress and believing in the impact of advanced learning technologies on learning outcomes, the conclusion has been reached that technology in itself does not solve the issues that are essential to education. Technology-enhanced learning is not a wonder wheel, where no other efforts are needed other than to bring the requisite technologies into educational settings. It is a continuously changing process, where different technologies are used to support students in their learning to become smart, motivated learners who know how to construct their knowledge and are supported by competent educators, who carry out predictive analyses of processes to use all the potential of the digital world. Learning in a digital environment provides access to information anytime, anywhere, but it is necessary to equip students with the competencies to analyse information and to construct their knowledge from a prism of its value, and there is also an urgent need to develop new digital learning materials to enable students to take advantage of the digital environment in the processes of constructing their knowledge and developing certain competencies that are needed in the digital environment.

In the chapters of this book, the authors discuss the perspectives from which the information found in the digital environment should be analysed, how freely available content can be transferred to digital learning materials, how digital technologies can be incorporated into educational practice, and what this brings to the development of knowledge from different dimensions. There is an attempt to define the predictive analytical competencies needed for teachers to operate in a technology-enhanced environment and an analysis of how to use advanced learning technologies to address inclusive education issues by providing access to learning materials. The authors of the chapters come up with ideas for digital solutions in educational settings in different dimensions of preschool and school education.

Organization of the book

The book consists of eight chapters, in which authors try to define the educational and pedagogical concepts for learning in the digital environment. A brief description of each of the chapters follows.

In the chapter "The concept of smart pedagogy for learning in the digital world", Linda Daniela proposes ideas of perspectives to consider how to evaluate the enhancement of technology use, which aspects should be taken into account when organizing the technology-enhanced learning process, and what educators should be aware of when creating digital learning materials.

In the chapter "Going beyond digital literacy to develop computational thinking in K-12 education", Divya Menon, Sowmya BP, Margarida Romero, and Thierry Viéville present a literature review on studies conducted to teach computer programming and computational concepts to K-12 students using visual programming tools, unplugged activities, and educational robotics while evaluating how it can also help improve CT skills.

In the chapter "The role of algorithmic thinking development in the learning of elementary school pupils aged 10–13", Miroslava Černochová, Hasan Selcuk, and Milan Svoboda present a case study that aimed to identify changes in elementary school pupils' algorithmic thinking through programming in Scratch 2.0.

In the chapter "Digital educational games in primary education: Revisiting the results of the research projects of the ETiE initiative", Emmanuel Fokides summarizes and critically evaluates 11 projects involving the use of 3D digital games. It is argued that the use of digital educational games results in better learning gains and conceptual understanding.

In the chapter "Digital learning materials: Does content make the difference in a digital world?", Polyxeni Kaimara, Sofia-Maria Poulimenou, and Ioannis Deliyannis argue that the dynamics of new interactive media support meaningful personal learning through collaboration, social interaction, and shared knowledge building, and present a cultural heritage-oriented case study where they concentrate on contemporary transmedia learning approaches.

In the chapter "Unpacking digital play in children's play", Jennifer Wong-Powell provides ideas on how digital play can be incorporated into an early childhood school setting where the adult plays a significant role in providing developmentally appropriate and intentional opportunities for digital play to occur.

In the chapter "Materialities in innovative education: A focus on small Italian schools", Giuseppina Rita Jose Mangione and Samuele Calzone explore the way digitally mediated materialities may support teaching practice in specific environments such as small schools, whose educational context has a high risk of students dropping out and problems of weakness and inequality.

In the chapter "A proposal of smart pedagogy for the school: Educational curation as an active methodology", Marilene Santana dos Santos Garcia and Luciano Frontino de Medeiros propose materials and tools that are available for the educational curator that support the development of critical thinking, fake news analysis, and creative activities through active and smart methodologies that can be applied in a school.

I am very grateful to all the authors who put their efforts into the preparation of their chapters and who also served as peer reviewers for their colleagues' materials. All the chapters included in this book were double-blind peer-reviewed, and all the authors took part in the peer-reviewing process, during which they did not know who the author of any particular article was. This process ensured the possibility of blind reviewing as well as for them to openly express their opinions. Such a procedure also ensures the quality of the articles as well as the mutual synergy of research ideas.

I hope that the book will contribute to the field and open up new lines of research, new ideas, and new concepts.

Professor Linda Daniela
University of Latvia, Latvia

Chapter 1

Concept of smart pedagogy for learning in a digital world

Linda Daniela

Introduction

Historically, there have been situations involving many different myths, challenges, and societal expectations regarding the outcome of education. In earlier times, education and acquired knowledge, which was gathered by their ancestors, was exclusively accessible to a certain class of people. Gradually, education became more democratic and accessible, creating the foundation for various major breakthroughs and the industrial revolution, and now, we are facing a digital revolution. As early as 1965, Moore predicted that technology would become more affordable as, with the introduction of new solutions with a low cost, the cost of joining different networks would be low too (Moore, 1965). The start of the digital technology revolution is considered to have been in the year 1971, when Intel introduced its microprocessor (Chan et al., 2006). The next major milestone in the widespread use of technological solutions was the definition of the principles of the World Wide Web (WWW) in the 1980s and its result with the first written Web browser in 1990, the outcome of Berners-Lee and Cailliau's work (1990), and its opening for public use in 1993. Initially, these advances had a greater impact on the information and communication technology (ICT) industry, but it is now clear that digital solutions have entered the educational landscape and their potential is already being widely exploited, and there is no controversy about their usefulness. The use of digital technology in educational settings has been given various names, such as computer-supported collaborative learning (Stahl, 2006); computer-assisted instruction, computer-aided instruction, computer-based learning or computer-mediated learning, which are considered synonyms (De Bruyckere & Kirschner, 2019); ICT for education (Usluel et al., 2008); e-learning (Pituch & Lee, 2006; Selim, 2007); mobile learning (Sharples et al., 2009); distributed learning (Mühlenbrock et al., 1998); asynchronous learning (DeNeui & Dodge, 2006; Dziuban et al., 2007); networked learning (Jackson, 2004); ubiquitous learning (Rogers, 2006); and technology-enhanced learning (Mooij et al., 2014), which are used to describe the application of ICT to teaching and learning (Kirkwood & Price, 2014), to provide flexibility in the mode of learning (Zhu et al., 2016), and to describe learning activities or learning environments in which technology is used as a facilitator, as a support, or as a scaffold learning processes (Barak, 2007; Plesch et al., 2013).

However, not all actors in the educational environment perceive these innovations with open arms, and not all educators strive to try and implement all innovations at once in their professional activities. In the public sphere, the educational environment is sometimes condemned for not using technology more intensively and confidently to change the traditional learning process, but for educators, technology in its own right does not change the learning process, the content being offered, and the importance of the student's cognitive, sensory, and socio-emotional developmental context. In this chapter, the term technology-enhanced learning is used and described in the next section, which considers perspectives using which the enhancement generated by the use of technology can be analysed. The following is an insight into the smart pedagogical principles applicable to a digital learning environment.

Technology-enhanced learning as the context for smart pedagogy

It is clear that as technology advances, it is increasingly being used in the educational environment to change different dimensions of the educational system, and one of the dimensions of the education system is the learning process that has been using computer technology and communication network solutions (WWW) already for some time. We currently know many different ways in which Advanced Learning Technologies (ALTs) can be used and will inevitably continue to expand and evolve as new technological solutions emerge, so at this point in time, the term *technology-enhanced learning* (TEL), which is increasingly used in different parts of the world, also includes earlier terms that describe the use of technology in different ways. TEL's various aspects have been intensively analysed for decades. The use of this term also illustrates what is expected from the use of technology, i.e. it can be introduced and used to support learning (Guri-Rosenblit & Gros, 2011).

Technology in the learning process is believed to help students make their own decisions, develop and diversify knowledge and skills, create diverse learning environments, and allow students to self-evaluate and provide feedback (Green et al., 2005), as TEL promotes an active, inclusive, ubiquitous form of learning and the formation of learning communities (Barak, 2012, 2014; Rafaeli et al., 2004). TEL is a continuously changing process where different technologies are used in the learning process to support students to become smart, motivated learners who know how to construct their knowledge and are supported by competent educators who carry out predictive analyses of processes (Daniela, 2019).

However, there are also some critical aspects in the use of the term TEL, as the term is currently used in a wide variety of contexts and it is not always clear which of these contexts is more relevant. In an educational setting, when it comes to TEL, it is assumed that technology enhances something. Unfortunately, technology is often at the forefront, and comes at the expense of enhancing the actual learning. Kirkwood and Price even point out that TEL is more often considered synonymous with the use of technology and the improvement of technology infrastructure (2014), with less emphasis on the enhancement section that should be attributed to learning.

It has to be admitted that in cases where the technology itself is not at the forefront, knowledge is the next most important indicator of the success and failure of the learning process. Many world-wide studies measure educational success in quantitative terms, a traditionally accepted measure of quality in education. When it comes to the enhancement of learning in a technology-enhanced (TE) environment, it is important to remember that the learning process has different goals, from the development of cognitive processes to knowledge growth to what it means to live and work in a TE environment. There is research that demonstrates that learners' knowledge in TEL has not improved significantly from the perspective of quantifiable parameters (Lee et al., 2018; Saunders & Gale, 2012). There are articles where the authors try to say that the impact of using technology in the learning process is exaggerated (Naidoo & Raju, 2012), or even create a drop-out risk (McPhee & Soderstrom, 2012). There are other articles claiming that TEL improves learning (Chamizo-Gonzalez et al., 2015; Chen & Lan, 2013; Lansari et al., 2010; Laru et al., 2012; Rockinson-Szapkiw, 2012), but these articles do not provide conclusive research evidence (Daniela et al., 2017). While not denying the necessity of measuring knowledge and measuring changes in that knowledge, and speaking of the objectivity of such measurements in terms of quantitatively analysed data, it is also important to accept that there are a number of factors that influence the evidence of TEL:

- **In the educational environment, it is not possible to ensure that all variables affecting knowledge development are kept constant, as measured solely by the impact of technology.** A comparison between experimental and traditional learning groups is also almost impossible, because here too, it is impossible to ensure that everyone involved in the learning process is exactly the same, or to ensure that a technology-driven group receives exactly the same materials, in exactly the same way, as is the case with a non-technology-driven group, and vice versa. It is not possible to guarantee that students who do not have access to technology in the educational environment do not use them outside the educational environment either.
- **Extensive and long-term research on the impact of the use of technology on knowledge can make no sense** because technology advances so rapidly that the technologies and technological solutions being explored may already be outdated by the time all research findings are finalized and recommendations made.

These considerations point to the need to use ongoing assessment methods, such as learning analytics solutions, to measure the growth of knowledge in a TEL environment, and to develop a pedagogical activity model where teachers, using predictive analytics competency (Daniela, 2019), predict potential results without knowing them yet. This will take pedagogical work to a new dimension, which will be discussed in more detail in the section on smart pedagogy as a way of organizing pedagogical activities in the TEL environment.

Returning to the TEL process, in addition to measuring and analysing specific quantifiable knowledge, it is also important to analyse what the use of technology can bring from the perspective of the learning experience, how attitudes to learning are shaped, and the effects of time constraints where there is no direct supervision of the learning process by the educator. What influences the retention of knowledge in operative memory, how analytical processes evolve and how they develop, and how these developments improve our understanding of different processes?

Thus, TEL needs to be analysed from several dimensions and each must apply its own research methodology. The results achieved must also be analysed not only from a statically stable point of view, but from a future perspective and a consideration that technology has become our daily ally as it evolves, and when analysing TEL. It is important to remember that enhancement can be analysed from (i) *the perspective of knowledge growth*; (ii) *the perspective of knowledge acquisition*; (iii) *perspectives on knowledge accumulation*; and (iv) *perspectives on knowledge access* (see Figure 1.1). It is not always possible to focus on one aspect of knowledge by analysing exactly what a TEL process has improved; the most important outcome may not be the increase in knowledge, but rather a change in the way knowledge is acquired, accumulated, and used to understand how the technology world works and to pave the way for new advancements.

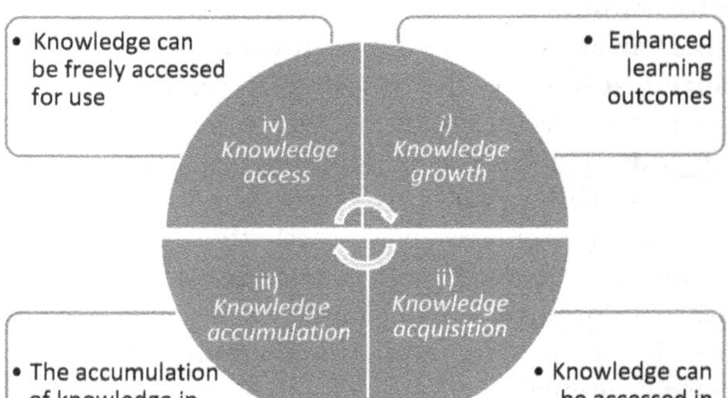

Figure 1.1 Knowledge dimensions.

Knowledge growth (i) is typically measured quantitatively across multiple dimensions, such as expansion, improvement, faster acquisition, a deeper understanding of some concepts, or other indicators of knowledge (hereafter, the generic term "knowledge gains" will be used to describe this dimension of knowledge growth).

From a **knowledge acquisition perspective (ii)**, technology allows access to knowledge in different ways and environments, and allows individuals to regulate the process of knowledge acquisition by themselves. These are knowledge and ideas that are acceptable both from (i) a scientifically verifiable perspective (the Earth is still round); (ii) a general public good perspective (nuclear power can also be destructive, even though it has been scientifically discovered and is commonly used); and (iii) the perspective of the rights and freedoms of the individual (the individual has the right to believe in anything, even the Flying Spaghetti Monster, as long as it does not endanger the general safety of society). It is important to remember that an open information space also contains knowledge that should be seriously evaluated and tested from all the above perspectives, and it is also the task of the pedagogical work to learn how to search for, evaluate, and accept as valuable and usable information that broadens the horizon with which the individual disagrees. Thus, in the digital environment, the knowledge that is freely available must also be accompanied by a sense of sincerity and ethical value.

The technology offering can also be analysed from the perspective of **knowledge accumulation (iii)**, as digital solutions allow everything to be digitized and accumulated in various ways, from simple solutions for material digitization to virtual museums that provide access to out-of-date exhibitions, or where tourist flows are not allowed to preserve cultural and historical values. How should the digital culture be preserved, how should history be digitized, how should these digital repositories be maintained and protected? Here, too, there are many questions about the interoperability of digital solutions, the accessibility of materials, and hence the prospect of **access to knowledge (iv)**. Not only the aspect of interoperability between different technologies, but also the principle of the openness of knowledge should be analysed here. Many research articles on the latest scientific discoveries are still not available to the general public, such as the Open Educational Resource. Thus, these findings can be accessed by representatives of well-funded universities/research institutions. Language barriers, socio-economic barriers, and special needs barriers all can create accessibility barriers to the worldwide community of knowledge builders.

In the following sections, ideas for a smart pedagogical process to operate in a technology-enhanced learning environment using digitalization capabilities will be presented.

Smart pedagogy – the driving force of TEL

Pedagogy is not just about what a teacher does, but about how learning happens, so what a teacher does affects what happens to students (Bruner, 1996). It is up to the teachers to take the lead in the educational process to offer ways in which students can access knowledge, as well as guidance on how to use it and how to help students develop certain competencies. With the advent of technology in the educational environment, Mishra and Koehler (2006) developed the concept of technological pedagogical and content knowledge (TPCK), defining what teachers need to know to work in a digital environment. The main criticism of this model is that it is pointed out that the mere fact that a teacher has a certain type of knowledge does not make him an expert in organizing a TEL process; the impact of the context is not clearly defined; different kinds of knowledge are not clearly defined; and the boundaries of that knowledge are not clearly shown (Pareto & Willermark, 2018). However, it cannot be denied that this model has made it possible to define the methodology of a teacher's activity in a TEL environment. Perhaps the reason for not fully exploiting the potential of technology and introducing fragments of TEL by testing some technological solutions, such as smart devices, learning management systems, or online materials, is that teachers are still not well prepared for the job with technology, and not all of them who had some training during the study process use any technologies in their educational work (OECD, 2019). Problems can also be related to the fact that technology is still changing and evolving, that there is not enough available content in the digital environment, and that not all teachers are ready to replace their pedagogical content knowledge (PCK) (Shulman, 1986), which has proven to be effective, with methods whose effectiveness has not yet been convincingly tested. On the one hand, you can see that learning is happening, but on the other hand, there are still unrealized expectations for the education system – miraculous, unseen achievements and assurances that there is finally an answer to the question of what needs to be done to make learning better, to reach higher achievements, faster progress, learning outcomes similar to those in Finland, Singapore, and Estonia. Can technology provide this, or on the contrary, does it only create a glimmer of light around the educational process but not add any value to the educational environment? These are issues that have been the focus of researchers for a long time.

Chan and colleagues predicted in 2006 that technological advances would ensure seamless learning when different devices enable learning without learning and when technological solutions are developed that allow everyone to use their own devices and change their learning styles, to access study materials, and to plan their own learning achievements (Chan et al., 2006). Seamless learning, self-regulated learning, and learning anywhere and anytime are terms that are often referred to in the TEL environment. When analysed from the perspective of the acquisition of specific technological skills, it really does take place and these skills are intuitively acquired. On the other hand, if one looks at the learning process from the point of view of improving one's learning achievements or developing one's cognitive processes, it has to be concluded that although more than a decade has passed since such statements were made, it is clear that it was an educational utopia. There are several reasons for this, but one of them is related to curriculum content in the digital environment: (i) There is a limited availability in the digital format; (ii) it does not involving interactivity; (iii) it is fragmented; (iv) it is too simplistic or too complicated; (v) it does not meet the specific learning needs of the individual student; and (vi) it does not include specific learning objectives to be achieved.

Bearing in mind the wide variety of technologies that are available, there is software that is available to work with and produce various materials, the content is available in a digital environment, and there may not seem to be any obstacles to creating appropriate material. However, these would be overly simplistic assumptions because: (i) The development of such materials requires both financial and time resources; (ii) specific expertise in the development of such materials; and (iii) knowledge of the use of such materials. This is also one of the reasons why the TPCK concept cannot be considered complete, as it involves knowledge of the use of materials but does not refer to the development of new materials and/or the transformation of freely available materials in learning content.

One of the solutions to supplement these ideas may be the concept of smart pedagogy. This term has recently appeared in the educational environment. It was initially mentioned in an article by Zhu and colleagues (2016), who stated that the TEL environment is to some extent considered a smart education environment where learning is intuitive, and who offered a four-tier architecture as a concept of classroom-based smart pedagogy: Differentiated instruction, followed by group-based collaborative learning, then individual-based personalized learning (predominantly interest-driven) and mass-based generative learning (predominantly through online interactions). Daniela (2019) gives a slightly different perspective on smart pedagogy, where a teacher uses all the available technologies to support students' learning, although these technologies should be evaluated from the perspective of learning outcomes that may be needed in the future. Spector says that the term 'smart' in relation to educational technology is usable if it is used effectively and efficiently (Spector, 2014), and the goal is to provide self-learning at the students' own pace and their ability to access personalized learning content according to their personal circumstances (Kim et al., 2013).

All of these goals are achievable if we look at them from the perspective of highly motivated students as they self-regulatedly learn because they are self-motivated. In the real world, it should be taken into account that students' prior knowledge is different, that their motivation for the cognitive load needed to acquire and operate new information metacognitively varies widely, and that the use of technology in the learning environment is not accompanied by smart pedagogical principles—not only content knowledge (CT), PCK (Shulman, 1986), and technological content knowledge (TCK) (Mishra & Koehler, 2006), but also predictive analytical competence (PAC) is needed (Daniela, 2019).

One of the reasons for defining pedagogical principles in a TE environment as *smart pedagogy* is based on the diagnostic method developed by IMB – Self-Monitoring Analysis and Reporting Technology, which included Predictive Failure Analysis (see https://www.computerhope.com/jargon/s/smart.htm). Thus, in the context of smart pedagogy, predictive analytical competence would mean that educators use their knowledge of human development, of the organizational regularities of the learning process, and of the opportunities offered by technology, and, at the same time, use that knowledge to forecast achievements to continuously analyse and improve ongoing learning processes. When thinking about the results to be achieved, one should keep in mind the different dimensions of knowledge, as well as the dimensions of socio-emotional development and physical development.

At present, many ideas that have been relevant to a specific time and stage of technology development and to the level of understanding of the place and role of technology in the educational process need to be reviewed and refined. It is clear that educators, who become the bearers of the idea of smart pedagogy and organize learning in a TE environment, have an important role to play, both by planning the growth of knowledge from the quantitative dimension and also by changing the qualitative dimensions of knowledge acquisition to prepare the next generation for life and work in the digital environment. The concept of smart pedagogy is formulated from the perspective of educator competence (Daniela, 2019), which outlines the principles to be followed in pedagogical work in a TEL process to promote the use of technology and digital solutions to promote learning, which many authors have already talked about by analysing the knowledge needed by educators. One suggestion is to develop predictive analytical competence, where the teacher predicts the result to be achieved through the use of materials and technologies he has not yet mastered and whose effectiveness has not yet been tested due to the difficulties encountered in obtaining evidence. In a TEL environment, it is essential for the teacher to predict the outcome to be achieved using certain criteria. The concept of smart pedagogy is triangular, in which the important cornerstones are:

1 **Human developmental regularities**, which include the conditions for the development of cognitive processes, the conditions for sensory development, as well as the conditions for socio-emotional development.
2 **The taxonomy of the educational process**, which includes the goals to be achieved and the regularities of the learning process needed to achieve these goals.

3 **Technological progress**, which entails the need for changes in teachers' pedagogical competence, of which one of the most important components is **predictive analytical competence** (Daniela, 2019).

These three sections cannot be analysed as independent from each other as their boundaries are blurred and overlap, because when thinking about the taxonomy of learning, it is important to know human developmental regularities, the regularities of the development of cognitive processes, to offer students tasks according to their abilities and zone of proximal development (Vygotsky, 1978). It is also important to know how to organize the learning process to motivate students to learn, and this depends on their level of cognitive development, interest, and readiness for cognitive load. Technology can help develop motivation, an interest in learning, and an understanding of more complex concepts of knowledge, and technologies can help avoid cognitive load. Analysing this further, we can assume that avoiding cognitive load can release cognitive capacity for the acquisition of other knowledge, and the fact that technology helps us perform complex mathematical activities, for example, cannot be considered as a negative effect on cognitive processes because cognitions occurs in searching for solutions. Conversely, if a student is confronted with a complex task rather than thinking about how to solve it, instead of seeking technology to solve it, he or she switches to another activity that does not require cognitive load, then the impact of technology may be seen as negative (see Figure 1.2) as cognitive processes do not progress to higher levels. If the knowledge is mechanically acquired, the ability to analyse and synthesize information and to create new concepts may be affected. As a result, individuals can be influenced by information that is freely available in the digital environment (fake news, propaganda, etc.) without questioning and analysing it. In this way, we return to the idea above that not all knowledge is considered valuable and necessary.

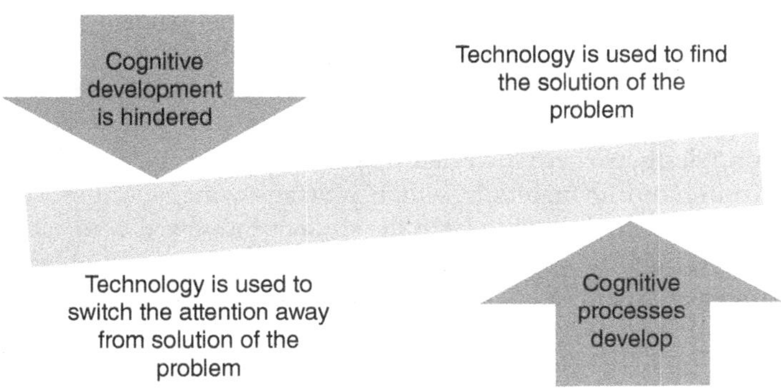

Figure 1.2 Challenge of the use of technology for cognitive development.

One of the tasks of smart pedagogy to ensure that technology is used to promote development is to maintain a balance in cognitive development, and there are several aspects to consider in particular:

i **Learning anywhere, anytime.** It is important to design and use learning materials that will ensure that attention to the need for cognitive load is not switched off and that the student will be ready to learn even when the learning is done by using digitally available materials. So they need to involving interactivity. By anticipating that learning can take place outside the teacher's field of influence, smart pedagogy's task is to teach students to analyse freely available information and how to build their own knowledge structures.

ii **Fascination with new technologies**. Assuming that most students are interested in everything new, the task of smart pedagogy is to stimulate this natural interest, while at the same time predicting the impact of the use of new technologies on students' cognitive, emotional, sensory, and physical development to promote the use of this fascination with technology and not merely as a source of interest. It is important to remember the relationship between interest and attention retention, where an overstimulation of interest during the developmental stage of cognitive processes may contribute to the development of a short-term attention span, which impedes the development of memory processes, which in turn are necessary for analysing and synthesizing information (Daniela, 2019).

Another task of smart pedagogy is related to the available technology and software, online opportunities for transforming teaching materials, and developing new teaching materials and assessing their impact on learning outcomes. So there must be a continuous:

i Process of learning analytics, which requires continuous monitoring of the learning process, analysing the data to predict student progress, seeing potential problems (Siemens, 2013; Zhu & Shen, 2013), and changing the learning process before problems occur, thus developing predictive analytical competence.

ii Learning design, where the learning process is designed to include learning through online materials, which are transformable to meet the needs of each student, balanced to keep the student interested while at the same time providing and stimulating learning, keeping the student focused, enabling assessment, etc.

Everything sounds so simple that it seems easy to do, but that is not the case, because when adopting new ideas one has to keep in mind that smart pedagogy is not tantamount to simply spicing the lesson up with attractive methods or digital media. It rather involves the transformation of the teacher's mental awareness – understanding the new perception of the teacher's and the student's roles and rejecting traditional thinking about the teaching process (stationary transmission of information) (Borawska-Kalbarczyk et al., 2019); it also includes significant time, finance, and new knowledge resources. In this educational environment, educators become not only skilful users of their PCK, not only technology users and predictors of achievements, but also orchestrants, when teachers manage in real-time in a context of multi-layered activities and multiple constraints (Dillenbourg, 2013).

As artificial intelligence solutions evolve, technology is predicted to provide each individual with exactly the information they need at the time to provide the most appropriate learning scenarios for their individual needs (Zhu et al., 2016). At the moment, it is difficult to predict what achievements artificial education will bring to the educational environment. Certainly, there will be solutions that will help people to learn, solutions that will be able to make our daily lives easier by doing things that can be automated and thus free up space and time for cognitive load.

Some final thoughts

Looking ahead to the future development of the educational environment, there are some issues that need further analysis:

A **Boundary blurring between formal and nonformal learning environments** – assuming that a student can study anywhere and anytime, as long as all barriers to access to teaching materials are eliminated and teaching materials are designed to promote self-directed learning, formal education could be considered to be likely to disappear. But is this really the case if we consider that self-directed learning is not a phenomenon that is inherited by everyone? It has to do with the peculiarities of human development, prior knowledge, attitude to learning, and motivation to learn, all of which are complex, multi-layered concepts. Assuming that the formal learning environment loses its status, could it not lead to problems that we do not even think about, such as:

- The very deep gap between those sections of the public who have access to a variety of educational resources, as they require access not only to OER but also to the devices on which these materials can be used, and those who do not: There is a need for access to the World Wide Web. Despite the fact that the prices and accessibility of these devices are becoming cheaper and can be seen as becoming affordable, there are still and will continue to be people who cannot afford it.

- If one were to take the utopian idea that all technology would also become free and that the barriers to accessing materials were completely eliminated, can one really think that young children will be self-taught to learn the alphabet and learn to read (any knowledge can be mentioned here)? I do not think this will happen, because you need someone to give them this basic knowledge and help them understand what can be done next, once one has mastered the skill of reading. So here again, there will be a gap between those people who will provide this opportunity for their children because they believe in the progression of knowledge and those many others who will be unable to do so.

B **Free access to information** means having access not only to true and valid information, but also to various misconceptions, falsehoods, half-truths, fake news, and educational propaganda. This critical mass of information requires certain critical thinking competencies to be able to evaluate information and understand its value. Will they evolve in an online education environment? I assume that there might also be a risk of information manipulation here. Therefore, civic and moral education and critical thinking must be developed. All of this can be done in a technology-enhanced learning environment, but it must be a somewhat driven process.

C **Data protection, data processing, privacy, and copyrights** are concepts that also need to be considered in an educational setting. How should children be taught to protect their data to avoid becoming victims of manipulation or crime? Does devising new rules for access to social networks help to develop these competencies? Does the fact that a child will be able to officially use any social network when they reach the age set by legal norms make them a skilled social network user? Let me say that age itself does not guarantee certain knowledge and competencies. So this is also a field where there is still much to be done from a pedagogical point of view to prepare children for this environment (Snapchat, Instagram, etc.), both from a data protection and an information analysis point of view.

D **The right to be yourself** – there are researchers who predict that technology will help to create a learning environment in which artificial intelligence will provide the individual with the exact knowledge they need at a given moment, offering the most appropriate form of learning. Does the individual have the right to decide what to learn, when to learn, how to learn? What will the conditions be under which these learning paths are offered? Will it be possible to disconnect this information to avoid its influence? For example, will a person be targeted for a specific area of science or industry?

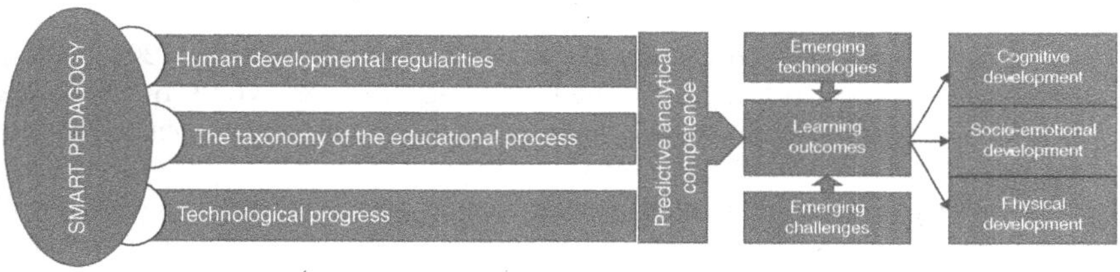

Figure 1.3 Flow of smart pedagogy.

This is certainly not an exhaustive list of future challenges that will need to be addressed, but they already require the attention of scientists, both in terms of technology-enhanced learning and in terms of smart pedagogical principles, to avoid a situation where teachers, using predictive analytical competency and seeing future potential problems, decide that there are more risks than opportunities, and therefore have to distance themselves from the technological environment. This is certainly not the answer to technological advances.

The use of technology in the learning process has enormous potential, both from the different dimensions of knowledge acquisition and from what technological solutions can bring to the improvement of the life of every individual. By developing smart pedagogical principles, TEL can take place not only in form but also in substance, which in turn will provide the basis for new achievements and innovations, and will help to develop smart services.

Overall, the main challenge for the educational environment today is not to stop at what is already known by incorporating current knowledge, but to help to see the unknown and ensure that the learning outcome is fully supported, taking into account opportunities for technological advancement that are not yet known, and to aware of the challenges that the technological environment poses to our own misconceptions about the technological environment itself (see Figure 1.3).

References

Barak, M. (2007). Transitions from traditional to ICT-enhanced learning environments in undergraduate chemistry courses. *Computers & Education, 48*(1), 30–43.

Barak, M. (2012). Distance education: Towards an organizational and cultural change in higher education. *Journal of Enterprising Communities: People and Places in the Global Economy, 6*(2), 124–137.

Barak, M. (2014). Closing the gap between attitudes and perceptions about ICT-enhanced learning among pre-service STEM teachers. *Journal of Science Education and Technology, 23*(1), 1–14.

Berners-Lee, T., & Cailliau, R. (1990). WorldWideWeb: Proposal for a HyperText Project. Retrieved 12 July 2019.

Borawska-Kalbarczyk, K., Tołwińska, B., & Korzeniecka-Bondar, A. (2019). From smart teaching to smart learning in the fast-changing digital world. In L. Daniela (Ed.), *Didactics of smart pedagogy: Smart pedagogy for technology enhanced learning* (pp. 23–40). New York, NY: Springer.

Bruner, J. (1996). *The culture of education*. Cambridge, MA: Harvard University Press.

Chamizo-Gonzalez, J., Cano-Montero, E. I., Urquia-Grande, E., & Muñoz-Colomina, C. I. (2015). Educational data mining for improving learning outcomes in teaching accounting within higher education. *International Journal of Information & Learning Technology, 32*(5), 272–285.

Chan, T. W., Roschelle, J., Hsi, S., Sharples, M., Brown, T., Patton, C., … Soloway, E. (2006). One-to-one technology-enhanced learning: An opportunity for global research collaboration. *Research and Practice in Technology Enhanced Learning, 1*(1), 3–29.

Chen, T., & Lan, Y. (2013). Using a personal response system as an on-class assessment tool in the teaching of basic college chemistry. *Australasian Journal of Educational Technology, 29*(1), 32–40.

Daniela, L. (2019). Smart pedagogy for technology enhanced learning. In L. Daniela (Ed.), *Didactics of smart pedagogy: Smart pedagogy for technology enhanced learning* (pp. 3–22). Cham: Springer.

Daniela, L., Kalniņa, D., & Strods, R. (2017). An overview on effectiveness of technology-enhanced learning (TEL). *International Journal of Knowledge Society Research, 8*(1), 79–91.

De Bruyckere, P., & Kirschner, P. A. (2019). Computer-assisted learning. In A. Tatnall (Ed.), *Encyclopedia of education and information technologies*. Cham: Springer. doi:10.1007/978-3-319-60013-0_73-1

DeNeui, D. L., & Dodge, T. L. (2006). Asynchronous learning networks and student outcomes: The utility of online learning components in hybrid courses. *Journal of Instructional Psychology, 33*(4), 256–259.

Dillenbourg, P. (2013). Design for classroom orchestration. *Computers & Education, 69*, 485–492.

Dziuban, C., Moskal, P., Brophy, J., & Shea, P. (2007). Student satisfaction with asynchronous learning. *Journal of Asynchronous Learning Networks, 11*(1), 87–95.

Green, H., Facer, K., Rudd, T., Dillon, P., & Humphreys, P. (2005). *Personalisation and digital technologies*. Bristol: Futurelab.

Guri-Rosenblit, S., & Gros, B. (2011). E-learning: Confusing terminology, research gaps and inherent challenges. *International Journal of E-Learning & Distance Education, 25*(1). Retrieved from http://ijede.ca/index.php/jde/article/view/729/1206

Jackson, D. (2004). Networked learning communities: Characteristics of 'networked learning' –What are we learning? Paper prepared for the International Congress for School Effectiveness and Improvement (ICSEI) conference, 'Building Bridges for Sustainable School Improvement', Rotterdam.

Kim, T., Cho, J. Y., & Lee, B. G. (2013). Evolution to smart learning in public education: A case study of Korean public education. In L. Tobias, R. Mikko, L. Mart, & T. Arthur (Eds.), *Open and social technologies for networked learning* (pp. 170–178). Berlin, Heidelberg: Springer.

Kirkwood, A., & Price, L. (2014). Technology-enhanced learning and teaching in higher education: What is 'enhanced' and how do we know? A critical literature review. *Learning, Media and Technology, 39*(1), 6–36.

Lansari, A., Tubaishat, A., & Al-Rawi, A. (2010). Using a learning management system to foster independent learning in an outcome-based university: A gulf perspective. *Issues in Informing Science & Information Technology, 7*, 73–87.

Laru, J., Naykki, P., & Jarvela, S. (2012). Supporting small-group learning using multiple Web 2.0 tools: A case study in the higher education context. *Internet And Higher Education, 15*(1), 29–38.

Lee, D., Morrone, A. S., & Siering, G. (2018). From swimming pool to collaborative learning studio: Pedagogy, space, and technology in a large active learning classroom. *Educational Technology Research and Development, 66*(1), 95–127.

McPhee, I., & Soderstrom, T. (2012). Distance, online and campus higher education: Reflections on learning outcomes. *Campus-Wide Information Systems, 29*(3), 144–155.

Mishra, P., & Koehler, M. (2006). Technological pedagogical content knowledge: A framework for teacher knowledge. *Teachers College Record, 108*(6), 1017–1054.

Mooij, T., Steffens, K., & Andrade, M. S. (2014). Self-regulated and technology-enhanced learning: A European perspective. *European Educational Research Journal, 13*(5), 519–528.

Moore, G. (1965). Cramming more components onto integrated circuits. *Electronics, 38*(8), 114–117. Retrieved from https://newsroom.intel.com/wp-content/uploads/sites/11/2018/05/moores-law-electronics.pdf

Mühlenbrock, M., Tewissen, F., & Hoppe, H. U. (1998). A framework system for intelligent support in open distributed learning environments. *International Journal of Artificial Intelligence in Education, 9*, 256–274.

Naidoo, S., & Raju, J. (2012). Impact of the digital divide on information literacy training in a higher education context. *South African Journal of Libraries & Information Science, 78*(1), 34–44.

OECD. (2019). ICT for teaching. In *TALIS 2018 Results (Volume I): Teachers and School Leaders as Lifelong Learners*. Paris: OECD Publishing. doi:10.1787/3ddd8e5a-en

Pareto, L., & Willermark, S. (2018). TPACK in situ: A design-based approach supporting professional development in practice. *Journal of Educational Computing Research*. doi:10.1177/0735633118783180

Pituch, K. A., & Lee, Y. K. (2006). The influence of system characteristics on e-learning use. *Computers & Education, 47*(2), 222–244.

Plesch, C., Kaendler, C., Rummel, N., Wiedmann, M., & Spada, H. (2013). Identifying areas of tension in the field of technology-enhanced learning: Results of an international Delphi study. *Computers & Education, 65*, 92–105.

Rafaeli, S., Barak, M., Dan-Gur, Y., & Toch, E. (2004). QSIAea web-based environment for learning, assessing and knowledge sharing in communities. *Computers & Education, 43*(3), 273–289.

Rockinson-Szapkiw, A. J. (2012). Should online doctoral instructors adopt audio feedback as an instructional strategy? preliminary evidence. *International Journal of Doctoral Studies, 7*, 245–258.

Rogers, Y. (2006). Moving on from Weiser's vision of calm computing: Engaging ubicomp experiences. In P. Dourish, & A. Friday (Eds.), *UbiComp 2006: Ubiquitous computing*. Lecture notes in computer science, vol. 4206. Berlin, Heidelberg: Springer.

Saunders, F. C., & Gale, A. W. (2012). Digital or didactic: Using learning technology to confront the challenges of large cohort teaching. *British Journal of Educational Technology, 43*(6), 847–858.

Selim, H. M. (2007). Critical success factors for e-learning acceptance. Confirmatory factor models. *Computers & Education, 49*(2), 396–413.

Sharples, M., Inmaculada, A. S., Marcelo, M., & Giasemi, V. (2009). Mobile learning: Small devices, big issues. In N. Balacheff, S. Ludvigsen, T. de Jong, & S. Barnes (Eds.), *Technology enhanced learning: Principles and products* (pp. 233–249). Heidelberg: Springer.

Shulman, L. S. (1986). Those who understand: Knowledge growth in teaching. *Educational Researcher, 15*(2), 4–14.

Siemens, G. (2013). Learning analytics: The emergence of a discipline. *American Behavioral Scientist, 57*(10), 1380–1400.

Spector, J. M. (2014). Conceptualizing the emerging field of smart learning environments. *Smart Learning Environments, 1*(1), 1–10.

Stahl, G. (2006). *Group cognition: Computer support for building collaborative knowledge*. Cambridge, MA: MIT Press. Retrieved from http://gerrystahl.net/mit/stahl%20group%20cognition.pdf

Usluel, Y. K., Askar, P., & Bas, T. (2008). A structural equation model for ICT usage in higher education. *Journal of Educational Technology & Society, 11*, 262–273.

Vygotsky, L. S. (1978). *Mind in society: The development of higher mental processes* (M. Cole, V. John-Steiner, S. Scribner, & E. Soubeman, Eds.). Cambridge, MA: Harvard University Press.

Zhu, Z. T., & Shen, D. M. (2013). Learning analytics: The science power of smart education. *E-education Research, 5*, 5–12.

Zhu, Z. T., Yu, M. H., & Riezebos, P. (2016). A research framework of smart education. *Smart Learning Environment, 3*, Article No. 4. doi:10.1186/s40561-016-0026-2

Chapter 2

Going beyond digital literacy to develop computational thinking in K-12 education

Divya Menon, Sowmya BP, Margarida Romero, and Thierry Viéville

Introduction

Across different OECD educational systems worldwide, K-12 education has transformed from being a textbook-based methodology to learning-focused pedagogy. To develop a learner-centred perspective, certain technology-enhanced learning (TEL) practices have been implemented for the development of competency-based education aiming to go beyond cognitive learning objectives (Motschnig-Pitrik & Standl, 2013). In this context, the selected TEL practices aim to empower learners with a creative and socio-critical perspective of digital technologies (Resnick, 2017; Romero et al., 2017b). In some cases, a learner-centred TEL approach has been facilitated by the possibilities of digital learning, along with learning resources and activities. The personalisation and creation of digital-inspired pedagogical concepts could provide learners with better access to and control over their learning experience. It could also provide educators with enhanced participation with subjects and a better awareness of learners' activities through learning analytic approaches (Siemens & Long, 2011).

When analysing the development of TEL approaches from a learning sciences perspective, certain studies focus on creating and using materials and tools in digital learning (Pachler et al., 2010), while others consider TEL from a creative perspective or as a way to develop computational thinking (CT) competency and a more enlightened approach to digital citizenship (Israel et al., 2015). This chapter goes beyond the idea of using existing digital learning materials (DLMs) and tools, to consider how learners can understand the underlying concepts of digital culture through the development of their CT competency in terms of formal systems (programming languages) and physical systems (sensors, actuators, and connectors, etc.) which interconnect as software and hardware.

Learning to code and CT development is a growing trend internationally (Chalmers, 2018; Faber et al., 2017). While it was earlier regarded as a suitable skill for advanced learners aiming for an IT career, today it is seen as an essential 21st-century skill that can be taught even to elementary students

(Grover & Pea, 2013). From this perspective, projects such as Class'Code, a French initiative to introduce programming in schools, have developed training for educators, resources, and an interdisciplinary learning community (Canellas et al., 2016). At the international level, numerous experiments have taken place in different contexts, although their systematic analyses have not been conducted. To support these learnings, this chapter will provide a reading and synthesis of different studies produced in educational science. Before introducing three different approaches in the way K-12 learners are introduced to CT, the concept's definition and a few worldwide initiatives to introduce programming at K-12 will be presented in the next section.

From coding to the development of CT competency

Coding could be learned as a technological skill in which a set of instructions are organised into a program. When learning to code, learners are acquiring knowledge about some concepts and procedures through which they develop an awareness about formal coding system, but they do not develop CT competency as described by Wing (2008), who defines it as "solving problems, designing systems, and understanding human behavior, by drawing on the concept fundamental to computer science" (p. 33). When we engage learners beyond learning to code, we should provide them with problem situations so that they get to analyse the situation, select the type of code and physical systems to use and develop an iterative process of improvement (Figure 2.1).

CT engages components related to the analysis of the problem situation and the way subjects organise and model the problem (problem analysis axis),

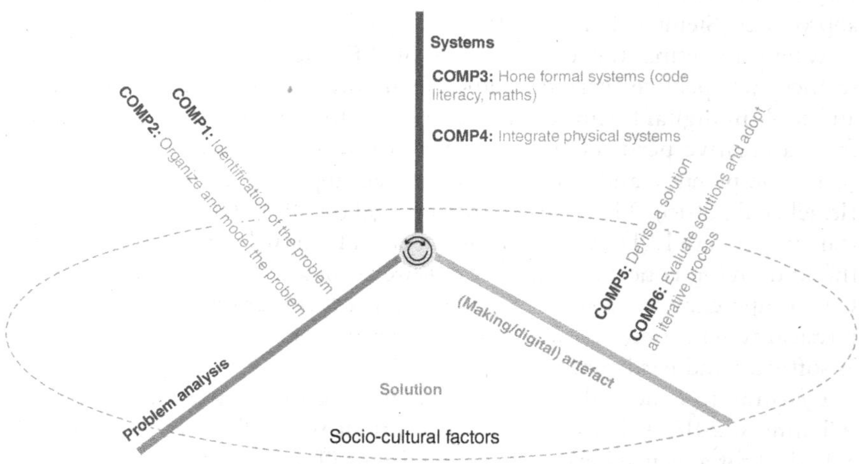

Figure 2.1 CT components (Romero et al., 2017a).

honing formal systems with the use of a certain programming language and integration of physical systems (systems axis) and devices of an intermediate solution, its evaluation and improvement (creation axis). When learners are only engaged in coding, they develop knowledge related to systems, but they do not engage in the full process of analysis, modelling, and iterative creation of a solution (Romero et al., 2017a).

Worldwide initiatives to introduce programming at K-12

Computer programming was first attempted to be integrated with the school curriculum in the 1980s (Popat & Starkey, 2019), but it started making an impact from 2010. Today, countries are adopting varied strategies when introducing coding to different learners. Chalmers (2018) mentions a research study conducted on four Australian primary school teachers to include coding and robotics in their classrooms through LEGO® WeDo® 2.0 kits. New College Worcester, England conducted a beta test with children with visual disabilities through Code Jumper, developed by Microsoft (Peters, 2019), a tool that uses "pods" or blocks with various shapes that can be connected into patterns and provides audio outputs. In another project by the British library's Young Researchers programme, two boys (aged 13–14 years) created a game in Mission Maker based on the poem *Beowulf* (Paula et al., 2018). The purpose of this activity was to foster learner creativity while integrating problem-solving skills with knowledge in arts and humanities through CT. Vaidyanathan (2013) also talks about a Bring-Your-Own-Device session she conducted with teachers at Los Altos School District in the USA using a "show-and-tell" method to emphasise that coding could be easily understood using everyday devices. Another innovative medium is used by Earsketch.gatech.edu, a virtual programming environment, that teaches coding in Python and JavaScript through music composing and runs annual competitions for interested students in the USA. There are others like KidsWhoKode.org, an Indian non-profit organisation, who follow a blended learning approach to teach coding to children aged 10–14 years from economically disadvantaged sections of society. They combine unplugged activities, online learning, classroom training, along with guest lectures and industry visits to make learners digitally empowered for the future (Kumari, 2018). In a very similar way, Kids Code Jeunesse in Canada supports teachers and educational professionals in developing CT skills (Romero et al., 2016a). Another example from India demonstrates the educational value and socio-cultural usage of CT as a way of life (Chandran, 2016; Dharavidiary.org). After being mentored and tutored by a filmmaker, a group of adolescent girls have learned to code using a donated laptop, created an app for the betterment of their community, and succeeded in expanding the project to more than 400 learners.

Through these different worldwide initiatives, we can observe the different approaches used by students to learn to code in schools and informal settings. The literature references for this chapter were acquired from sources such as ScienceDirect.com and ResearchGate.net primarily from the years 2006 till current date. Special emphasis was laid on collecting information with real data analysis regarding the progress made in the teaching of computer programming and on CT using digital pedagogical methods. In the next section, we will describe the three approaches identified from these initiatives and a literature review on the way kids learn programming in schools.

Diversity in the way kids learn programming in schools

While continuing to grow at a fast pace, CT education also exhibits advanced and innovative implementation methods (Heintz et al., 2016), and educational sciences can now work on this object of study. We consider three approaches that are used to introduce kids to programming at the school level: Unplugged computing, visual programming tools, and educational robotics.

The first approach focuses on learning CT through *unplugged computing*. Cicirello (2013) describes unplugged activities in computer science as a way of teaching CT concepts with hands-on tools and activities rather than digital devices. Some studies are now interested in how CT can be imparted without the use of technology. They focus on how unplugged activities use pedagogy-based methods that help reduce the mental block most learners seem to have about code learning (Grover et al., 2018). Unplugged activities can help demystify CT for young and old learners. For example, KidsWhoKode.org conducts unplugged activities to start every coding session with games such as "Simon Says" to introduce conditional actions, get learners to write their daily activities to help identify sequences and loops, or use arts and crafts or music composition to understand the basics of programming. Brackmann et al. (2017) also describe how methods, such as decomposition activity (breaking down real-world tasks into required steps), map activity (using four-directional arrow keys to move objects from point A to point B on a map), children's song activity (converting a song into an algorithm to identify variables, repetition, and conditionals), etc., were tried on students aged 10–12 years from Madrid to introduce CT concepts. Most researchers seem to agree that unplugged activities can help learners understand these concepts (Bell et al., 2009; Moreno-León & Robles, 2015a).

The second approach considers using *visual programming tools*, with an intent to develop creative computing skills, rather than just a "software developer". Since their introduction as a way to improve CT skills (Brennan & Resnick, 2012; Wilson & Moffat, 2010), a variety of programming tools, such as Scratch, Hopscotch, and Kodable have come up in recent years. Among these, studies are focused primarily on Scratch (Moreno-León & Robles,

2015a; Resnick & Siegel, 2015) and how it is used. Rose et al. (2017) conducted a study with a group of 40 children aged 6 and 7 years on whether programming tools can help develop CT skills in young learners. They used ScratchJr, a version of Scratch designed for younger learners (aged 5–7 years), and Lightbot, an educational puzzle that helps learners program a robot using block-based instruction. Since Lightbot only contains a subset of commands compared to Scratch and doesn't allow learners to explore the tool freely, the study concluded that Scratch provides learners with more freedom to experiment, thus leading to improved problem-solving ability. Similarly, Baker (2017) provides a comparison of features and capabilities between Scratch and Blockly. It appears that there are features provided by Blockly (including the option to translate programming blocks directly into other languages, such as JavaScript, PHP, and Python) which might be lacking in the current version of Scratch. On the other hand, Scratch provides additional features, such as the ability to share code for completed projects within the user community (which would lead to a sense of contributing, knowledge sharing, and self-directed learning), and the ease of using coding to create animations (which might easily attract and engage young learners).

The third approach is *educational robotics*, which offers the additional potential to learn how to interact with a connected or programmed object. Educational robotics not only allows improving skills regarding formal systems (component 3) but also manipulating and understanding physical systems (component 4). Several studies available establish the feasibility of making robotic devices with children. They show some virtues and limits too. Using robotics in educational activities seems to engage learners more than many of the other instructional methods (Papastergiou, 2009) because they get to play an active role in the learning process while using their creativity and problem-solving ability, thus increasing learner motivation. According to Carbonaro et al. (2004), learners get the opportunity to imbibe knowledge better since they get to collaborate with and showcase their robotic projects to a wider audience. In the process, learners also seem to acquire additional skills. For example, Benitti (2012) mentions that apart from helping learners understand robotics, it can also help improve skills, such as problem-solving, logical thinking, and scientific inquiry. Castro et al. (2018) highlight another interesting observation that in a sample of 389 students selected to learn educational robotics, the success level of both girls and boys appeared to be the same. But while a few (Schopler & Toplis, 2008; Whittier & Robinson, 2007) experiment with using robotics to integrate subjects that are not related to physics and mathematics, most of the learning with educational robotics revolves around robotic concepts, such as "robot programming, robot construction, artificial intelligence, algorithm development and mechatronics" (Mitnik et al., 2008, p. 1). Some of the other possible drawbacks of educational robotics are covered by McNally et al. (2006), who categorise them into logistical and pedagogical disadvantages. Logistical disadvantages cover issues of cost and accessibility (it can only be offered to learners in limited numbers and controlled environments) and pedagogical disadvantages focus on the limited concepts that are currently being covered through robotics-based education. Benitti (2012) also mentions that educational robotic activities predominantly use different versions of LEGO robots that are designed for children who are aged 7 years and older. It often appears that because of the 'fun' nature of these educational activities, they are mostly conducted as after-school or summer-camp activities, and hence, might lack the quantitative data required to understand how educational robotics can help with the learning of subjects other than STEM better.

For an analysis on how learning sciences has approached CT (Barr & Stephenson 2011; Wing 2008), we will review studies in each of these three approaches to learn programming in K-12. The present review cannot claim to be exhaustive but aims to be representative of available knowledge. Because of a bias that favours these learnings, special attention has been paid to the few 'critical' studies that also show limitations and disadvantages of these approaches. Meta-analyses or reviews of only studies with real factual results have been considered here, rather than "testimonials" as explained by Hickmott et al. (2017). It should be noted that all studies are cautious, including the earliest studies in education science on Logo programming language proposed by Papert (Clements & Meredith, 1992).

Approach 1: Unplugged activities to develop CT competencies

Until a few years ago, there seemed to be a lack of interest among students to learn computer science. This could be attributed to a few factors, including the assumed complexity in understanding computer science, the 'let's jump into programming' learning approach used by most educators, and the fact that it was regarded by most learners as a highly intellectual and therefore unattainable skill and something only suitable for male learners (Bell et al., 2009; Taub et al., 2012). But with the persistent integration of unplugged activities, these assumptions seemed to be finally changing for a majority of learners (Bell et al., 1998; Bell et al., 2009). During unplugged or computer-less activities, children learn concepts at the heart of computing in general or robotics in particular through play. Unplugged activities have been observed to be effective for learning CT (Brackmann et al., 2017).

According to Tricot (2017), computer thinking skills include knowing how to break down a problem into simpler subproblems, thinking about tasks to be done to solve a problem in terms of steps and actions (algorithm), describing problems and solutions at different levels of abstraction, which makes it possible to identify similarities between problems and subsequently to be able to reuse solution elements. Such improvement of these skills is confirmed by Moreno-León and Robles (2015b) on about 60 pupils aged 9–11 years, after a study, including an initiation into programming turned into a learning of algorithmic thought, with a key element: These teachers had just been trained in this new learning, and this was conducted in an English learning class. To establish these CT skills, Moreno-León et al. (2015a) link each competency to the use of Scratch language constructions. They then propose a tool (Dr.Scratch.org) to assess the level of competence by using such constructions from the framework proposed by Brennan and Resnick (2012) included in a manual, such as Huseyin et al. (2017) to name only the most recent. In these studies, the leverage is in the shift from procedural learning of programming to multi-disciplinary integration of creative programming, as explained by Romero (2016a). An example of skill transfer between learning to programme and other learning is given by the significant improvement of the left–right distinction after a robot activity (Romero et al., 2016a), and while other types of transfers are fairly likely, they are yet to be established. This positive effect on the ability to solve problems and, to a lesser extent, on reasoning and spatialisation is established on groups of pupils at the end of primary and middle school (Román-González et al., 2017). These results are consistent with the work of Chen et al. (2017), which include the use of educational robotics. The competency assessment on this subject has been finely studied by Lepage and Romero (2017) and Romero et al. (2017a). An evaluation grid tool is used to assess and validate participants' engagement in solving problems of a certain complexity and authenticity in a critical, empathetic, and creative way while using strategies and processes of computer science to create one or more solutions.

Approach 2: Visual programming or learning to programme with Scratch

Scratch uses constructivist pedagogy that encourages learners to work with an artifact and tinker with it while completing a task, thus using that as a basis for building knowledge. Resnick et al. (2009) mention that most Scratch users belong to the age group of 8–16 years, though there are a lot of adult users too. They also define the purpose of Scratch as a medium not aimed at creating professional programmers, but one that could enable learners to use programming skills to interpret real-world situations creatively and systematically, while "using programming to express their ideas" (p. 1).

Wilson and Moffat (2010) studied 20 children aged 8 years in disadvantaged areas of Glasgow and observed a positive experience through learning with Scratch to overcome frustrations and anxieties of learning, including children who learn well despite having difficulties. Gulbahar and Kalelioglu (2014) observed primary school youth for whom learning Scratch can increase programming skills, but with no increase in problem-solving skills in general, unless the authors' literature review shows it to design specific problem-solving activities from Scratch. Also, the use of Scratch on a group of college freshmen compared to traditional methods of learning on paper is not superior to the level of the results (Tekerek & Altan, 2014), but at the level of pleasure to learn and in generating student engagement. This study also shows the lack of difference in performance between girls and boys. The commitment of students is widely used in the studies of learning with Scratch (Ortiz-Colon & Romo, 2016) on a college-level class, even though the evaluation results are too scattered to lead to a definitive conclusion. What seems well established is Scratch's didactic value in terms of computer literacy as reported by Korkmaz (2016), wherein about 50 university entrance students were able to learn a language of professional programming, with this learning being significantly improved by prior use of Scratch. It has also been observed (Fesakis & Serafeim, 2009) that Scratch helped some post-baccalaureate students increase their self-confidence in digital tools and led to a positive influence on their digital uses in education. At the interface between programming learning and "tangible" computing, Horn et al. (2009) show that introducing programming using tangible objects (like a building set) through family workshops in a museum helped increase the performance of learners. This result is to be compared to what Boissel (2017) obtained with a tangible system for school initiation to programming for sighted and visually impaired children. Another study concerns the learning of programming with tangible objects and Correll et al. (2014) show the positive contribution of the commitment to studying the sciences through questionnaires. It should also be noted that Scratch can lend itself to pedagogical approaches that are engaging for the child, such as the Scratch Community Blocks system, which allows children to analyse their learning data (Dasgupta & Hill, 2017). Learning to programme in Scratch is sometimes coupled with unplugged computing activities, tangible devices, or participatory approaches, and in its broader conditions, there is a positive differentiation.

Approach 3: Educational robotics or usefulness of playful and educational robotics

Let us first consider available studies that analyse the feasibility of introducing educational robotics even to very young children. Sklar et al. (2003) attribute this to the ease with which "off-the-shelf robotic kits, such as LEGO Mindstorms, FischerTechnik Mobile Robot, and Elekit SoccerRobo" can be bought and introduced into classrooms (p. 5). As before, robotic activities are attractive but do not systematically offer leverage to help with other learning, especially if we are left with the acquisition of the technological know-how. Gordon et al. (2015) conducted a study on how the use of a robot to help children learn something promotes/encourages their curiosity. The systematic literature review in educational robotics of Benitti (2012) simplifies our analysis as it shows that real improvements in learning are observed, but this is not always the case, the differentiating factor being the methodological (teacher's attitude and adapted workspace), and curriculum (which skills are targeted). The danger seems to be to make robotics for robotics, to remain at the level of a work of copying technology. Few studies look, beforehand, if the handling of the robotic system is easy, contrary to Desprez et al. (2018) who take the trouble to validate the usability of their device. This study also shows that out of 70 publications analysed, only around 10 or so include a real quantified study beyond statements a priori, or occasional testimonials, such as Robinson (2005), which are not devoid of interest, but whose validity, in general, is questionable. More recently, Kim et al. (2015) show how robotic experimentation is used to teach and learn science, technology, engineering, and mathematics, linking these disciplines, and resulting in the motivational commitment of the teacher. The link between educational robotics and CT (see next section) is made by Atmatzidou and Demetriadis (2016) who note the need for long-time learning, with consistency in quality of performance regardless of age or gender, with the fact that girls take longer to complete a process (we note here an overinterpretation of the authors who refer to girl students and say that "they need more time to achieve the same result" while facts show "that they spend more time", and if the time had been constrained there is no proof that the results would have been worse). In a smaller study, Bers et al. (2014) establish the capacity of about 50 children (kindergarten level) to appropriate the first notions of robotics adapted to their age. A qualitative observation (Sullivan et al., 2013) confirms the feasibility of introducing the youngest (in this case, pre-kindergarten level) to educational robotics. More specifically, based on interviews, Highfield et al. (2008) show learning in spatialisation and geometry from the manipulation of a programmable object and activity involving the body. More recently Spolaôr and Benitti (2017) make a systematic literature review of the use of educational robotics for other learning and conclude on the real potential, established in several cases, especially when the approach is combined with a constructivist pedagogical approach. In his work on IniRobot resources with Thymio-II, Roy (2015) shows a real increase of skills on a group of 24 primary-level children with a final score of 93% at the post-test establishing that learning is accessible to everyone. The educational work with Thymio-II demonstrating its ability to engage primary and middle school children was established by the creators of the robot (Riedo et al., 2013), including the ability to appropriate basics of robotics (Magnenat et al., 2012).

Discussion

Table 2.1 presents a comparison of features, benefits, and limitations between the three approaches used in developing CT skills among K-12 learners. The review conducted in this paper shows that all three approaches have certain limitations as well as unique benefits. Educators need to take into account a few critical factors when considering the most suitable approach in their

Table 2.1 Features, advantages, and limitations of the three approaches to CT development

Comparison criteria	*Unplugged activities*	*Visual programming tools*	*Educational robotics*
Tools required	• Hands-on tools and activities • No digital tools • No internet	• Digital software • Internet	• Educational robot (usually a programmable toy) • Digital software • Internet
Suitability	Suitable for learners of all ages	Suitable for learners aged seven years and above	Suitable for learners aged seven years and above
Prior knowledge required	No prior knowledge of programming or computers required	Requires basic knowledge about computers from a user perspective	Requires basic knowledge about programmable interfaces (such as Thymio or Beebot)
Investment required	Very inexpensive compared to other approaches	Requires investment in computers	Requires investment in educational robots (and computers, if software is required)
CT components covered	All CT components could be introduced, some of them in a metaphorical way (physical systems – hardware)	All CT components except physical systems (hardware), but there is often a focus on formal systems (code literacy)	All CT components could be covered
Benefits	• Useful in introducing initial coding concepts • Could help in reducing inhibitions that learners might have about code learning • Can be used anywhere with any material available (free resources for activities available online)	• Focuses on developing creative computing skills • Programming tools, such as Scratch, are free • Has the potential to be a fun and engaging method for learners • Learners get real-time feedback through a visual representation of their activities • Can be used to integrate STEM subjects	Could offer the following benefits: • Provide advanced learning about coding and hardware • Allow learners to display their creation to others • Help improve other skills, such as problem-solving, logical thinking, and scientific inquiry
Limitations	• Doesn't offer any experience with hardware or software • Doesn't provide hands-on practice in coding • Unless learning objectives are briefed, learners might not understand the concepts being represented through the unplugged activities	• Even if software is free (e.g. Scratch), computers have a cost that might not be suitable for learning environments with limited monetary resources • Doesn't cover physical systems • While learners may be able to reuse or modify existing codes, they might be unable to write own codes	• Involves higher cost than other approaches • Can only be offered to a few learners at a time • Most of the learning centres around robotic concepts

educational environment. The first and primary limitation faced by most educators is in terms of availability of resources and existing infrastructure. For schools from economically disadvantaged sections of society, it may be a challenge to use the educational robotics approach, since they require more investment. In such cases, using unplugged activities or visual programming tools would be more suitable for their purpose. One also needs to consider the CT competencies that are to be integrated and the level of expertise that the educator wants learners to master. From the mentioned examples, it appears that while unplugged activities might be a good approach to introduce CT competencies to learners, visual programming tools help learners understand them and practice their use in creative ways, and educational robotics provide them with an advanced learning experience.

By learning how to "use computers in the solution of life problems for production purposes" (Korkmaz et al., 2017) learners, particularly school children, will gain something essential regarding their education, and much more. That is why CT initiation is beyond learning to programme or control robots. This statement has to be supported with concrete proof, and the previous literature review brings some positive answers to this statement. Indeed, for this statement to be meaningful, the concept of CT must be well defined. ISTE (2015) defines CT as the common reflection of creativity, algorithmic thinking, critical thinking, problem-solving, cooperative thinking, and communication skills. When these skills are taken into consideration together, they explain a brand-new thinking skill that is called CT (Korkmaz et al., 2017). When considering the potential of the three approaches to aid in developing CT skills that go beyond technical skills and integrate other essential skills mentioned above, it can be seen from the reviewed studies that all the three approaches appear to be promising.

What can and cannot be expected from CT learning

The present review tends to show that learning to code can promote autonomy and develop skills regarding problem-solving and research approach, but it is not automatic as it requires support in these two components. For instance, one track is to explain a solution to students to enable them to build the next one, such as a variant, and allowing them to take inspiration from existing solutions to be adapted, which could help reduce the large gap imposed by autonomous learning. Learning to code is considered by some authors as a relevant and effective skill that requires the know-how to manipulate. However, learning CT also requires notional knowledge learning, thus being cognitively active is very important. Group and project learning can address multiple skills, but can also fail, if not managed well. This learning can be fostered by helping students organise themselves, or by giving them roles and defining progressive stages. Such an approach makes it possible to engage pupils in the activities, and their perception *a posteriori* is very positive. Better engagement of students in their learning activities, thanks to the interest generated by CT activities, may be considered as the ultimate positive effect of such a paradigm.

Conclusion

When we go beyond simple learning of programming, for example, through unplugged computing activities or by promoting learning of algorithmic thinking, we can achieve a real positive effect at the primary level and beginning of college, including with newly trained teachers. The leverage here is the transition from procedural learning of programming to multi-disciplinary integration of creative programming. The studies covered in this chapter point to positive learning outcomes in terms of CT, but further research needs to be developed to analyse the learning transfer to other disciplines. The analysis of current practices in the way programming is introduced in K-12 education has led to a huge diversity in terms of pedagogical strategies.

In many cases, all three approaches are combined in practice, such as by performing an unplugged activity or manipulating a pedagogical robot, before learning programming. However, to our best knowledge, the combination of these three approaches has not been studied formally. To this end, future work is needed to design a paradigm to study how to combine these different approaches in an integrated environment to engage learners using unplugged activities along with other methods, such as by using tangible robotic programming tasks. We hope to be able to do this through an escape-game inspired experience. This is the goal of a future ANR #CreaMaker/AIDE project structured to observe the creative aspects of CT as well as to collect automatic logs that will track the components of CT competencies. Similarly, efforts in the emerging field of computational learning sciences are also combining selective loosely defined tasks within a context in which the analysis of sensors and logs would permit the integration of machine-learning approaches to first analyse user competencies and second to help learners develop their learning process further by providing cues at appropriate moments for external and internal learning regulations. These few results in the learning sciences seem to confirm the combined approach of unplugged computing activities, visual programming tools, and educational robotics projects such as Class'Code.

Acknowledgement

We aim to acknowledge the support of EducAzur cluster, INRIA, and the ANR #CreaMaker (ANR-18-CE38-0001) project for supporting the development of this chapter.

References

Atmatzidou, S., & Demetriadis, S. (2016). Advancing students' computational thinking skills through educational robotics: A study on age and gender relevant differences. *Robotics and Autonomous Systems*, *75*, 661–670.

Baker, S. (2017). Blockly VS Scratch: What's best for me? Retrieved 13 April 2019, from https://www.robotlab.com/blog/blockly-vs-scratch-whats-best-for-me

Barr, V., & Stephenson, C. (2011). Bringing computational thinking to K-12: What is involved and what is the role of the computer science education community? *ACM Inroads, 2*(1), 48.

Bell, T. C., Alexander, J., Freeman, I., & Grimley, M. (2009). Computer science unplugged: School students doing real computing without computers. Retrieved 12 April 2019, from ResearchGate website: https://www.researchgate.net/publication/266882704_Computer_Science_Unplugged_school_students_doing_real_computing_without_computers

Benitti, F. B. V. (2012). Exploring the educational potential of robotics in schools: A systematic review. *Computers & Education, 58*(3), 978–988.

Bers, M. U., Flannery, L., Kazakoff, E. R., & Sullivan, A. (2014). Computational thinking and tinkering: Exploration of an early childhood robotics curriculum. *Computers & Education, 72*, 145–157.

Boissel, S. (2017). Programmer les yeux fermés. Retrieved 5 April 2019, from binaire website: http://binaire.blog.lemonde.fr/2017/05/22/apprendre-a-programmer-les-yeux-fermes/

Brackmann, C. P., Roman-Gonzalez, M., Robles, G., & Moreno-Leon, J. (2017). Development of computational thinking skills through unplugged activities in primary school. In *Proceedings of the 12th Workshop on Primary and Secondary Computing Education*, 65–72. Nijmegen, Netherlands: ACM.

Brennan, K., & Resnick, M. (2012). New frameworks for studying and assessing the development of computational thinking. In *Proceedings of the 2012 Annual Meeting of the American Educational Research Association*, Vol. 1, p. 25. Vancouver, Canada.

Canellas, C., De La Higuera, C., Peinchaud, É., & Roche, M. (2016). Class' Code a un an... et c'est un commencement. *1024 – Bulletin de La Société Informatique de France*, (9), 19–24.

Carbonaro, M., Rex, M., & Chambers, J. (2004). Using LEGO robotics in a project-based learning environment. *The Interactive Multimedia Electronic Journal of Computer-Enhanced Learning, 6*(1), 55–70.

Castro, E., Cecchi, F., Valente, M., Buselli, E., Salvini, P., & Dario, P. (2018). Can educational robotics introduce young children to robotics and how can we measure it? Retrieved 12 April 2019, from ResearchGate website: www.researchgate.net/publication/327470707_Can_educational_robotics_introduce_young_children_to_robotics_and_how_can_we_measure_it

Chalmers, C. (2018). Robotics and computational thinking in primary school. *International Journal of Child-Computer Interaction, 17*, 93–100.

Chandran, R. (2016). Girls code in India to tackle challenges of slum living - Reuters. Retrieved 13 April 2019, from https://in.reuters.com/article/india-children-tech/girls-code-in-india-to-tackle-challenges-of-slum-living-idINKCN0YI20F

Chen, G., Shen, J., Barth-Cohen, L., Jiang, S., Huang, X., & Eltoukhy, M. (2017). Assessing elementary students' computational thinking in everyday reasoning and robotics programming. *Computers & Education, 109*, 162–175.

Cicirello, V. (2013). (PDF) A CS unplugged activity for the online classroom. Retrieved 12 April 2019, from ResearchGate website: https://www.researchgate.net/publication/244988711_A_CS_Unplugged_Activity_for_the_Online_Classroom

Clements, D. H., & Meredith, J. S. (1992). Research on logo: Effects and efficacy. *Journal of Computing in Childhood Education, 4*(4), 263–290.

Correll, N., Wailes, C., & Slaby, S. (2014). A one-hour curriculum to engage middle school students in robotics and computer science using cubelets. In H., Asama, T., Fukuda, T., Arai,, & I. Endo (Eds.). *Distributed Autonomous Robotic Systems* (pp. 165–176). Springer, Berlin, Heidelberg.

Dasgupta, S., & Hill, B. M. (2017). Scratch community blocks: Supporting children as data scientists. In *Proceedings of the 2017 CHI Conference on Human Factors in Computing Systems - CHI '17*, 3620–3631.

Desprez, T., Noirpoudre, S., Segonds, T., Caselli, D., Roy, D., & Oudeyer, P.-Y. (2018). *Design and evaluation of pedagogical robotic kits*. Retrieved from https://hal.archives-ouvertes.fr/hal-01688310

Dharavidiary.org. (n.d.). Retrieved 13 April 2019, from https://www.dharavidiary.org/

Dr.Scratch.org. (n.d.). Retrieved 4 April 2019, from http://www.drscratch.org/

earsketch.gatech.edu. (n.d.). Retrieved 9 April 2019, from http://earsketch.gatech.edu/landing/#/

Faber, H. H., Wierdsma, M. D., Doornbos, R. P., van der Ven, J. S., & de Vette, K. (2017). Teaching computational thinking to primary school students via unplugged programming lessons. *Journal of the European Teacher Education Network, 12*, 13–24.

Fesakis, G., & Serafeim, K. (2009). Influence of the familiarization with 'Scratch' on future teachers' opinions and attitudes about programming and ICT in education. In *Proceedings of the 14th Annual ACM SIGCSE Conference on Innovation and Technology in Computer Science Education*, 258–262.

Gordon, G., Breazeal, C., & Engel, S. (2015). Can children catch curiosity from a social robot? *Proceedings of the Tenth Annual ACM/IEEE International Conference on Human-Robot Interaction - HRI '15*, 91–98.

Grover, S., Jackiw, N., Lundh, P., & Basu, S. (2018). (PDF) Combining non-programming activities with programming for introducing foundational computing concepts. Retrieved 12 April 2019, from ResearchGate website: https://www.researchgate.net/publication/326539412_Combining_Non-Programming_Activities_with_Programming_for_Introducing_Foundational_Computing_Concepts

Grover, S., & Pea, R. (2013). Computational thinking in K–12 a review of the state of the field. *Educational Researcher, 42*(1), 38–43.

Gulbahar, Y., & Kalelioglu, F. (2014). (PDF) The effects of teaching programming via Scratch on problem solving skills: A discussion from learners' perspective. Retrieved 5 April 2019, from ResearchGate website: https://www.researchgate.net/publication/271746747_The_Effects_of_Teaching_Programming_via_Scratch_on_Problem_Solving_Skills_A_Discussion_from_Learners'_Perspective

Heintz, F., Mannila, L., & Färnqvist, T. (2016). A review of models for introducing computational thinking, computer science and computing in K-12 education. *Frontiers in Education Conference (FIE), 2016 IEEE*, 1–9. IEEE.

Hickmott, D., Prieto-Rodriguez, E., & Holmes, K. (2017). (PDF) A scoping review of studies on computational thinking in K–12 mathematics classrooms. Retrieved 11 April 2019, from ResearchGate website: https://www.researchgate.net/publication/321760448_A_Scoping_Review_of_Studies_on_Computational_Thinking_in_K-12_Mathematics_Classrooms

Highfield, K., Mulligan, J., & Hedberg, J. (2008). Early mathematics learning through exploration with programmable toys. In *Proceedings of the Joint Meeting of PME, 32*, 169–176.

Horn, M. S., Solovey, E. T., Crouser, R. J., & Jacob, R. J. K. (2009). Comparing the use of tangible and graphical programming languages for informal science education. In *Proceedings of the 27th International Conference on Human Factors in Computing Systems - CHI 09*, 975.

Huseyin, O., Gary, W., & Tugba, O. H. (2017). *Teaching computational thinking in primary education*. IGI Global, Hershey, PA, USA.

Israel, M., Pearson, J. N., Tapia, T., Wherfel, Q. M., & Reese, G. (2015). Supporting all learners in school-wide computational thinking: A cross-case qualitative analysis. *Computers & Education, 82*, 263–279.

ISTE. (2015). Computational thinking leadership toolkit. Retrieved 5 April 2019, from https://id.iste.org/docs/ct-documents/ct-leadershipt-toolkit.pdf

KidsWhoKode.org. (n.d.). Retrieved 10 April 2019, from <KidsWhoKode /> website: https://www.kidswhokode.org/

Kim, C., Kim, D., Yuan, J., Hill, R. B., Doshi, P., & Thai, C. N. (2015). Robotics to promote elementary education pre-service teachers' STEM engagement, learning, and teaching. *Computers & Education, 91*, 14–31.

Korkmaz, Ö. (2016). The effects of Scratch-based game activities on students' attitudes, self-efficacy and academic achievement. *International Journal of Modern Education and Computer Science, 8*(1), 16–23.

Korkmaz, Ö., Çakir, R., & Özden, M. Y. (2017). A validity and reliability study of the computational thinking scales (CTS). *Computers in Human Behavior, 72*, 558–569.

Kumari, B. (2018, April 9). KidsWhoKode aims to be a student-driven movement: Code of small things [News]. Retrieved 10 April 2019, from Bangalore Mirror website: https://bangaloremirror.indiatimes.com/bangalore/cover-story/kidswhokode-aims-to-be-a-student-driven-movement-code-of-small-things/articleshow/63672465.cms

Lepage, A., & Romero, M. (2017). Évaluation par compétences d'activités de programmation créative avec l'outil #5c21. Retrieved 5 April 2019, from ResearchGate website: https://www.researchgate.net/publication/320310837_Evaluation_par_competences_d'activites_de_programmation_creative_avec_l'outil_5c21

Magnenat, S., Riedo, F., Bonani, M., & Mondada, F. (2012). A programming workshop using the robot "Thymio II": The effect on the understanding by children. In *2012 IEEE Workshop on Advanced Robotics and Its Social Impacts (ARSO)*, 24–29.

McNally, M., Goldweber, M., Fagin, B., & Klassner, F. (2006). Do LEGO Mindstorms robots have a future in CS education? In *Proceedings of the 37th SIGCSE Technical Symposium on Computer Science Education - SIGCSE '06*, 61.

Mitnik, R., Nussbaum, M., & Soto, A. (2008). (PDF) An autonomous educational mobile robot mediator. Retrieved 10 April 2019, from ResearchGate website: https://www.researchgate.net/publication/200744655_An_autonomous_educational_mobile_robot_mediator

Moreno-León, J., & Robles, G. (2015a). Dr. Scratch: A web tool to automatically evaluate Scratch projects. In *Proceedings of the Workshop in Primary and Secondary Computing Education*, 132–133. ACM.

Moreno-León, J., & Robles, G. (2015b). Computer programming as an educational tool in the English classroom: A preliminary study. *2015 IEEE Global Engineering Education Conference (EDUCON)*, 961–966. IEEE.

Moreno-León, J., Robles, G., & Roman-Gonzalez, M. (2015). (PDF) Dr. Scratch: Automatic analysis of Scratch projects to assess and foster computational thinking. Retrieved 5 April 2019, from ResearchGate website: www.researchgate.net/publication/281714025_Dr_Scratch_Automatic_Analysis_of_Scratch_Projects_to_Assess_and_Foster_Computational_Thinking

Moreno León, J., Robles, G., & Román-González, M. (2016). Code to learn: Where does it belong in the K-12 curriculum? *Journal of Information Technology Education: Research, 15*, 283–303.

Motschnig-Pitrik, R., & Standl, B. (2013). Person-centered technology enhanced learning: Dimensions of added value. *Computers in Human Behavior, 29*(2), 401–409.

Ortiz-Colon, A. M., & Maroto Romo, J. L. (2016). Teaching with Scratch in compulsory secondary education. *International Journal of Emerging Technologies in Learning (IJET), 11*(02), 67.

Pachler, N., Cook, J., & Bachmair, B. (2010). Appropriation of mobile cultural resources for learning. *International Journal of Mobile and Blended Learning, 2*(1), 1–21.

Papastergiou, M. (2009). Exploring the potential of computer and video games for health and physical education: A literature review. *Computers & Education, 53*(3), 603–622.

Paula, B. H., Burn, A., Noss, R., & Valente, J. A. (2018). Playing Beowulf: Bridging computational thinking, arts and literature through game-making. *International Journal of Child-Computer Interaction, 16*, 39–46.

Peters, A. (2019, January 22). These tactile blocks teach blind kids to code. Retrieved 8 April 2019, from Fast Company website: https://www.fastcompany.com/90294418/these-tactile-blocks-teach-blind-kids-to-code

Popat, S., & Starkey, L. (2019). Learning to code or coding to learn? A systematic review. *Computers & Education, 128*, 365–376. doi:10.1016/j.compedu.2018.10.005

Resnick, M. (2017). Fulfilling Papert's dream: Computational fluency for all. In *Proceedings of the 2017 ACM SIGCSE Technical Symposium on Computer Science Education*, 5–5. ACM.

Resnick, M., Maloney, J., Monroy-Hernández, A., Rusk, N., Eastmond, E., Brennan, K., … Kafai, Y. (2009). Scratch: Programming for all. *Communications of the ACM, 52*(11), 60–67.

Resnick, M., & Siegel, D. (2015, November 10). A different approach to coding. Retrieved from Bright website: https://medium.com/bright/a-different-approach-to-coding-d679b06d83a#.b9jcw2js5

Riedo, F., Chevalier, M., Magnenat, S., & Mondada, F. (2013). Thymio II, a robot that grows wiser with children. In *2013 IEEE Workshop on Advanced Robotics and its Social Impacts*, 187–193. IEEE.

Robinson, M. (2005). Robotics-driven activities: Can they improve middle school science learning? *Bulletin of Science, Technology & Society, 25*(1), 73–84.

Román-González, M., Pérez-González, J.-C., & Jiménez-Fernández, Ç. (2017). Which cognitive abilities underlie computational thinking? Criterion validity of the computational thinking test. *Computers in Human Behavior, 72*, 678–691.

Romero, M. (2016a, May 5). La pensée informatique. Retrieved 5 April 2019, from #CoCreaTIC website: https://margaridaromero.wordpress.com/2016/05/05/la-pensee-informatique/

Romero, M. (2016b, May 25). De l'apprentissage procédural de la programmation à l'intégration interdisciplinaire de la programmation créative. Retrieved 5 April 2019, from #CoCreaTIC website: https://margaridaromero.wordpress.com/2016/05/25/de-lapprentissage-procedural-de-la-programmation-a-lintegration-interdisciplinaire-de-la-programmation-creative/

Romero, M., Davidson, A.-L., Cucinelli, G., Ouellet, H., & Arthur, K. (2016a). Learning to code: From procedural puzzle-based games to creative programming. *Learning and Teaching Innovation Impacts*. Presented at the International Conference on University Teaching and Innovation, Barcelona, Spain.

Romero, M., Dupont, V., & Pazgon, E. (2016b). À gauche ou à droite du robot? Test de perspective décentrée gauche-droite par le biais d'une activité sur papier et d'une activité de robotique pédagogique. Retrieved 5 April 2019, from CIRTA website: http://www.cirta.org/index.php/50-banque-de-textes-actes-colloque-2016/221-a-gauche-ou-a-droite-du-robot-test-de-perspective-decentree-gauche-droite-par-le-biais-d-une-activite-sur-papier-et-d-une-activite-de-robotique-pedagogique

Romero, M., Lepage, A., & Lille, B. (2017a). Computational thinking development through creative programming in higher education. *International Journal of Educational Technology in Higher Education, 14*(1), 42.

Romero, M., Lille, B., & Patino, A. (Eds.). (2017b). *Usages créatifs du numérique pour l'apprentissage au XXIe siècle* (Vol. 1). Québec: Presses de l'Université du Québec.

Rose, S. P., Habgood, M. P. J., & Sheffield, T. J. (2017). (PDF) An exploration of the role of visual programming tools in the development of young children's computational thinking. Retrieved from www.ejel.org/issue/download html?idArticle=602

Roy, D. (2015). *Optimisation des parcours d'apprentissage à l'aide des technologies numériques* (PhD thesis, CNAM, Paris). Retrieved from https://tel.archives-ouvertes.fr/tel-01252695/document

Schopler, E., & Toplis, R. (2008). Journal of autism and developmental disorders. Retrieved 10 April 2019, from ResearchGate website: https://www.researchgate.net/publication/314067414_Journal_of_Autism_and_Developmental_Disorders

Siemens, G., & Long, P. (2011). Penetrating the fog: Analytics in learning and education. *EDUCAUSE Review, 46*(5), 30.

Sklar, E., Eguchi, A., & Johnson, J. (2002, June). RoboCupJunior: learning with educational robotics. In G. A., Kaminka, P. U., Lima, & R. Rojas, (Eds.). *Robot Soccer World Cup* (pp. 238-253). Springer, Berlin, Heidelberg.

Spolaôr, N., & Benitti, F. B. V. (2017). Robotics applications grounded in learning theories on tertiary education: A systematic review. *Computers & Education, 112*, 97–107.

Sullivan, A. R. Kazakoff, E., & Umashi Bers, M. (2013). The wheels on the bot go round and round: Robotics curriculum in pre-kindergarten. *Journal of Information Technology Education: Innovations in Practice, 12*, 203–219.

Taub, R., Armoni, M., & Ben-Ari, M. (2012). (PDF) CS unplugged and middle-school students' views, attitudes, and intentions regarding CS. Retrieved 12 April 2019, from ResearchGate website: https://www.researchgate.net/publication/241623893_CS_Unplugged_and_Middle-School_Students%27_Views_Attitudes_and_Intentions_Regarding_CS

Tekerek, M., & Altan, T. (2014). The Effect of scratch environment on student's achievement in teaching algorithm. *Online Submission, 6*(2), 132–138.

Tricot, A. (2017). L'innovation pédagogique : Mythes et réalités. Retrieved 5 April 2019, from http://www.cafepedagogique.net/LEXPRESSO/Pages/2017/09/05092017Article636401937485284884.aspx

Vaidyanathan, S. (2013, December 9). We need coding in schools, but where are the teachers? - EdSurge News. Retrieved 9 April 2019, from EdSurge website: https://www.edsurge.com/news/2013-12-09-opinion-we-need-coding-in-schools-but-where-are-the-teachers

Whittier, E., & Robinson, M. (2007). Teaching evolution to non-English proficient students by using Lego robotics. Retrieved 10 April 2019, from ResearchGate website: https://www.researchgate.net/publication/234707215_Teaching_Evolution_to_Non-English_Proficient_Students_by_Using_Lego_Robotics

Wilson, A., & Moffat, D. C. (2010, September). Evaluating scratch to introduce younger schoolchildren to programming. *PPIG* (p. 7). Retrieved from Scratched website: http://scratched.gse.harvard.edu/sites/default/files/wilson-moffat-ppig2010-final.pdf

Wing, J. M. (2008). Computational thinking and thinking about computing. *Philosophical Transactions of the Royal Society of London A: Mathematical, Physical and Engineering Sciences, 366*(1881), 3717–3725.

Chapter 3

The role of algorithmic thinking development in the learning of elementary school pupils aged 10–13

Miroslava Černochová, Hasan Selcuk, and Milan Svoboda

Introduction

Over the last decade, a great deal of attention has been given to computational thinking (CT) by scholars worldwide. The concept of CT was used by Seymour Papert (1980), and since then (as highlighted by Grover & Pea, 2013), CT has been discussed by various scholars in respect of its definition, planning, and implementation in teaching and evaluation. CT can be seen as "taking an approach to solving problems, designing systems and understanding human behaviour that draws on concepts" (Wing, 2006, p. 33). Moreover, Weintrop et al. (2016, p. 127) defined CT in the context of mathematics and science education as "the form of a taxonomy consisting of four main categories: Data practices, modelling and simulation practices, computational problem-solving practices, and systems thinking practices". Aho (2012, p. 832) described CT as "the thought process involved in formulating problems so their solutions can be represented as computational steps and algorithms".

In this study, our focus is concerned with algorithmic thinking (AT). Being an important part of computational thinking skill, AT can be defined as the ability to understand, implement, assess, and design algorithms to solve a range of problems. Algorithmic design as an outcome of AT can be understood as a process of "creating an ordered series of instructions for solving similar problems or for performing a task" (Hsu et al., 2018, p. 299). As stressed by Hromkovič and Lacher (2017, p. 4)

> already Euclid formulated algorithms in his 'Elements', among them being the famous Euclid's algorithm. The term 'algorithm' is due back to al-Khwarizmi, who wrote a book about Indian digits around the year 825. Humankind tried to develop algorithms as long as anyone can remember. People generated knowledge in order to understand the world around them and specially to apply this knowledge to develop 'procedures' to reach their goals

Fellows et al. (2005 in Tekerek & Altan, 2014, p. 133) described algorithm as "a set of instructions that are required to follow for completing a task".

Furthermore, Grover and Pea (2018) argued that algorithms and AT, together with patterns and pattern recognition, abstraction and generalization, evaluation, and automation, are the key components of CT, whereas Hromkovič et al. (2017) claimed that even though AT and CT were introduced in different eras and consider these two concepts to be equivalent and interchangeable, at times, they can be expressed as the same concept. For Futschek (2006), AT is an ability that is necessary at any stage of the problem-solving process.

In the Czech Republic, educational blocks-based programming environment Scratch and educational robotics (e.g., LEGOMindstorms, ozobots) are being introduced into schools to develop pupils' computational thinking following the idea that pupils would "become not merely passive consumers of computer technology but rather active designers of computer programs and applications" (Dewey, 2017).

The aim of this case study is to identify changes in Czech elementary school pupils' (grade 5 and 6) AT through programming in Scratch 2.0.

Literature review

Some computer scientists are not convinced that CT enhances general cognitive skills (e.g. problem-solving skills). Denning (2017, p. 37), declares that, "many education researchers have searched for supporting evidence about it but have not found any". The findings of Weintrop and Wilensky's (2017, p. 32) from a five-week, quasi-experimental study in a high school, which focused on comparing isomorphic block-based and text-based programming environments, showed that in both groups, pupils improved their scores between pre-and post-tests. Nevertheless, pupils in the block-based group showed greater learning gains and higher levels of interest. Rogalski and Samurcay (1990) declare that learning to program in any language is not an easy task. Programming teachers would agree with Rogalski and Samurcay that they are well aware of the myriad difficulties that beset beginners. Concerning the characteristics of the beginner or novice programmers, Lahtinen et al. (2005, p. 14) stated that, "novice programmers are typically limited to surface knowledge of programs. They often approach programming 'line by line' rather than using meaningful program structures". Moreover, a study by Robins et al. (2003) focused on identifying problems experienced by novice programmers found the following three key issues: (1) Relating to basic program design, (2) algorithmic complexity in certain language features, and (3) fragility of novice knowledge.

It has been commonly agreed upon by computer scientists that "Scratch is amazing on many levels. It's a real programming language that people use in dozens of countries to do programming that they find personally meaningful and useful" (Guzdial, 2008, p. 64). As discussed by Maloney et al. (2010) and Durak (2018), teaching programming with Scratch influences pupils' self-directed learning skills when working in online learning environments and reflective skills towards problem-solving. Scratch programming media and its visual design enables especially young pupils to express themselves easily and provide ways that facilitate their learning through programming (Durak, 2018).

To date, the studies carried out on AT in the primary school context are recent in the field. A study by Benton et al. (2018) aimed to explore the pupils' (aged 10–11) understanding of algorithm following their engagement with one year of ScratchMaths, a curriculum designed to develop computational and mathematical thinking skills through learning to program. About 161 pupils from six different English schools participated in the study and undertook a set of written tasks to assess their interpretations and evaluations of different algorithms. The findings of the Benton et al. study revealed that, through Scratch, pupils found algorithms such as simplification and abstraction intuitive or challenging. Meerbaum-Salant et al. (2013) investigated whether Scratch can be utilized to teach concepts of computer science. The authors developed new learning materials for middle-school pupils (grade 9, aged: 14–15). Participants of the study were administered from two classrooms: Classroom one consisted of 18 pupils (11 boys and 7 girls) and classroom two consisted of 28 pupils (all boys). The results showed that pupils could successfully learn important concepts of computer science. However, as Meerbaum-Salant et al. (2013, p. 261) highlighted, "difficulties were encountered in teaching specific topics, such as repeated execution, variables, and concurrency". These authors suggested that these problems could be overcome by modifications to the teaching process.

Beyond these studies, it is clear that further research is useful to understand better the process of pupil's algorithmic thinking development not only in mathematics and computer science education but also in other subjects leading to pupil's improvement in solving problems of everyday life.

The study

Aims, objectives, and research questions

The aim of this case study is to identify changes in Czech elementary school pupils' (grade 5 and 6) algorithmic thinking through programming in Scratch 2.0. We set the following four objectives and formulated two research questions:

1 to identify pupils' initial AT skills;
2 to design teaching of programming in Scratch for pupils and investigate the contribution of Scratch activities to AT development;
3 to ascertain the influence of programming (with Scratch) on pupils' AT;
4 to discover how or if programming influenced pupils' approach to learning.

RQ1: Can teaching in Scratch develop pupils' algorithmic thinking?
RQ2: Do algorithmic activities in Scratch influence pupils' engagement in programming?

Case study approach

The study was arranged as a case study with 48 Czech elementary school pupils (grades 5 and 6) within a compulsory subject, Information and Communication Technology (ICT). The current study identifies changes in pupils' algorithmic thinking through programming in Scratch across time in the ICT subject that consists of 22 lessons.

Participants and the setting

This case study took place between September 2017 and May 2018 in one typical elementary school in central Prague with about 250 pupils including foreign pupils from China or Ukraine. One of the authors, who taught the participants, obtained an official consent to implement the study in his classrooms from the school's management. All 48 pupils and their parents were informed about the study and their consent was taken. A "purposeful sampling" (Creswell, 2012) was used in this current study and two classrooms were selected from one of the classrooms of the researcher: One classroom in grade 5 and the second one in grade 6. Before the study began, all 48 participants (35.4% female, aged 10–13) were divided into four groups, A, B, C, and D, based on the availability and capacity of the computer lab (Table 3.1).

Pedagogical design

The schedule of the case study was divided into three parts with six phases (Table 3.2). The teaching approach to algorithmic thinking development was based on activities without computers (unplugged activities) and on programming in Scratch 2.0.

Table 3.1 Overview of participants (N = 48)

Class	*Group*	*Number of pupils (% female)*	*Label*
Grade 5	Group 1	11 (27.2)	A
	Group 2	12 (16.7)	B
Grade 6	Group 3	13 (38.5)	C
	Group 4	12 (58.3)	D

Table 3.2 Time schedule and aims of the case study

Part	*Phase*	*Date*	*Aim*	*Number of 45-minute lessons*	*Research method*
Preparatory part	Phase 1	September 2017	To obtain background information about participants	4 lessons	Questionnaire 1
	Phase 2	October 2017	To find out participants' initial algorithmic skills		Test A1
Pedagogical experiment	Phase 3	October 2017–February 2018	Learning Scratch	10 lessons	Test B1
	Phase 4	February–April 2018	To design the concluding Scratch project	8 lessons	DrScratch
Impact of the experiment	Phase 5	April 2018	To find out the impact of learning in Scratch on participants' algorithmic skills		Test A2
	Phase 6	May 2018	To obtain information from participants about their learning to program		Questionnaire 2

Methods

The data collection process of this study included two questionnaires and three tests (see Table 3.2), and an analytical evaluation using the application DrScratch, which can identify the usage of *logic constructs* (if; if else; logic operations), *parallelism* (two scripts on green flag; two scripts on key pressed; two scripts on sprite clicked on the same sprite, etc.), *synchronization*

Table 3.3 Research methods used during the case study

Research methods	*Description*
Questionnaire 1	Questionnaire 1, with 16 closed-ended and 9 open-ended questions, aimed to obtain information from participants' ICT background (ICT engagement, amount of time spending with digital technology, pupils' relationship to computer technology, the ways of using computer technology and the pupils' relationship to programming, etc.).
Test A1	Test A1 aimed to identify participants' algorithmic abilities before the pedagogical experiment. It consists of 16 test tasks (12 closed-ended and 4 open-ended). Two test tasks were divided into two parts that were rated separately. A maximum score a pupil could get was 18 points.
Analytic evaluation using the application DrScratch	*Topic of the concluding project:* Designing and programming of a computer game in Scratch. All 48 participants' projects in Scratch were analysed by using DrScratch.
Test B1	Test B1 aimed to measure the impact of teaching on participants' AT and participants' ability to do algorithm and to understand coding in Scratch. The test consisted of 17 tasks (12 closed-ended and 5 open-ended focused on producing flowcharts for Scratch scripts). A maximum score a pupil could get was 17 points. The data were collected through a printed test that the participants solved entered directly in the classroom on the lesson. Tasks in the test were rated separately.
Test A2	Test A2 aimed to measure the impact of programming in Scratch on participants' AT and to see a participants' progress. The test consisted of 16 tasks (12 closed-ended and 4 open-ended). Two test tasks were divided into two parts which were rated separately. A maximum score a pupil could achieve was 18 points. (Test A2 had the same set of test tasks but their order differed).
Questionnaire 2	Questionnaire 2, with 23 closed-ended and 3 open-ended questions, aimed to obtain information from participants about their learning to program.

(wait; broadcast, when I receive message; stop all; stop program; stop programs sprite, etc.), *user interactivity* (green flag; key pressed; sprite clicked; ask and wait; mouse blocks, etc.), *abstraction and problem decomposition* (more than one script and more than one sprite; definition of blocks; use of clones), *data representation* (modifiers of sprites properties; operations on variables; operations on lists), and *algorithm or flow control* (sequence of blocks; repeat, forever, etc.) (Moreno-León et al., 2015, p. 6). Paper-based questionnaires (1 and 2) and tests (A1, A2, and B1) were given to participants in the classroom during one lesson (45 minutes), see Table 3.3.

Findings

This section presents the findings from the data gathered from two questionnaires and three tests (A1, A2, and B1) and an analytical evaluation using DrScratch. To present the findings in a logical sequence, we have divided this section into four parts. Part 1 is based on data from questionnaire 1, and we provide some background information about participants' ICT use outside the school. Part 2 gives the analytical evaluation of programming with Scratch 2.0 using the test B1 and DrScratch. Part 3 presents and compares the results of tests A1 and A2. The final Part 4 presents the results of questionnaire 2.

Part 1: Participants

A majority of pupils use a computer at home, only two (4%) stated that they could not work with a computer outside school. The average number of days per week when pupils use a computer outside the school is six. Two-thirds of the pupils stated that their parents were limiting their time spent with a computer. A large number of pupils (42%) spend about three to four hours per day with a computer. On average, each pupil spends about 2 hours and 20 minutes per day with a computer outside the school.

Most parents check the time their children spend with a computer. Only 27% of pupils stated that parents also control a type of activity their child is doing on a computer. Out of school, the pupils use computers mainly playing computer games (34% of time spent with computers) and only 10% of the time by preparing and learning for school. Only 11% of pupils give time to creative activities with a computer (e.g., drawing and editing photos or videos).

In most cases, at least one parent uses a computer for his/her job (61%). Nine percent of pupils had a parent working as a programmer. About 11% of pupils own a programmable toy (Lego Mindstorm, programmable robots, etc.). Only 14% of pupils had previous experience of programming. In all, 10% of pupils said they had worked with GameMaker. No pupil had any experience with Scratch before the pedagogical experiment (Figure 3.1).

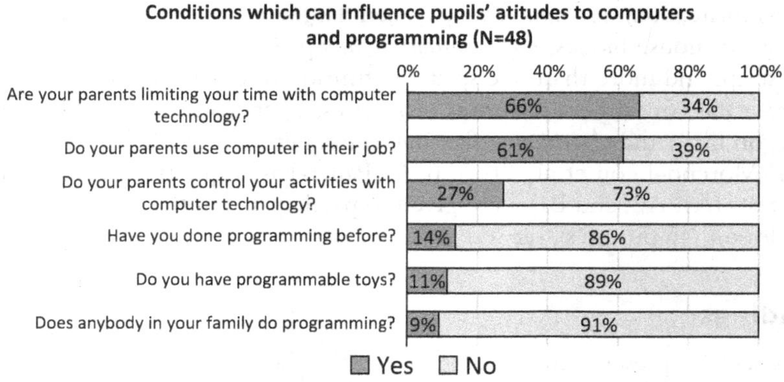

Figure 3.1 Conditions at pupils' homes for using a computer.
Source: M. Svoboda (2018), questionnaire I.

Part 2: Analytical evaluation of programming with Scratch using test B1 and DrScratch

This part focuses on a concept of algorithm when pupils did coding in Scratch. This was investigated using the test B1 and DrScratch. The tasks in test B1 required pupils to demonstrate their ability to read scripts, to interpret scenarios in Scratch, or to decide which of the codes can be used for programming a described process. In five tasks of test B1 pupils designed an algorithm for a Scratch scenario. All 17 tasks in test B1 were directly related to Scratch. The test B1 focused on understanding not only the concept of algorithm (in Scratch context), but also of cloning, variables, a cycle, events and messages, and parallelisms and sequences (Figure 3.2).

In all, 84% of pupils demonstrated how they understood the concept of algorithm in their activities with Scratch (Figure 3.3).

Each pupil has designed and programmed his/her own game in Scratch. All their outcomes were analysed using DrScratch that assigns one value from {0; 1; 2; 3} for each item (competence), see Figure 3.4. The lowest results were achieved in items of abstraction (respectively abstract thinking) and parallelism.

21 points is the maximum that pupils can achieve in DrScratch analysis. 54.2% of pupils gained at least 10 points in the DrScratch analysis.

Pupils achieved maximum points (3 points) in the DrScratch analysis mainly for two items *algorithm (flow control)* and *logic constructs* (see Figure 3.5).

According to DrScratch analysis, 80% of Scratch projects reach at least developing or master level (Figure 3.6).

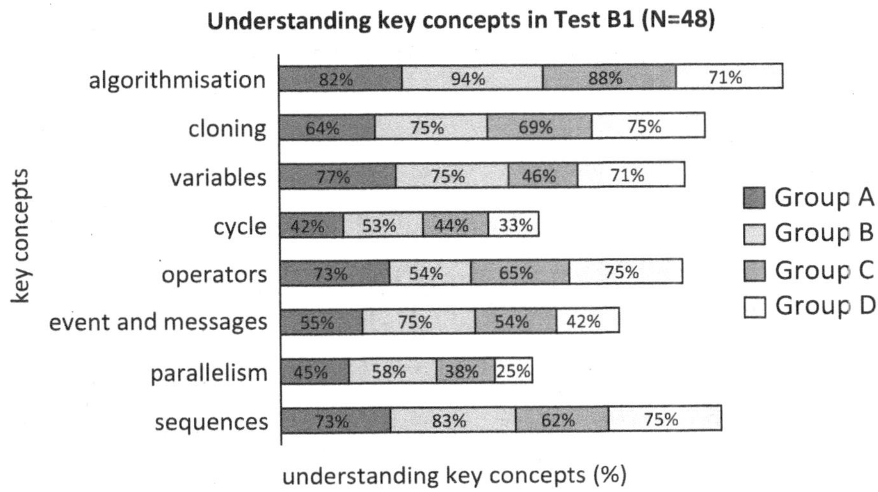

Figure 3.2 Pupil's understanding key concepts in test B1.
Source: M. Svoboda (2018).

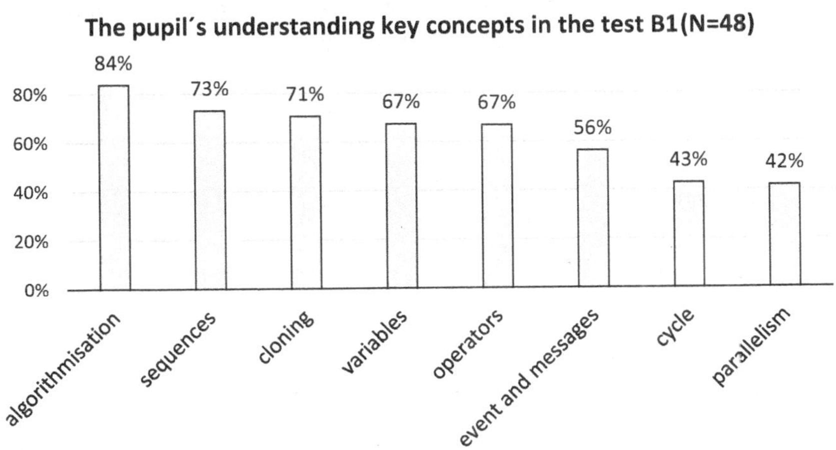

Figure 3.3 Pupil's understanding key concepts in test B1.
Source: M. Svoboda (2018).

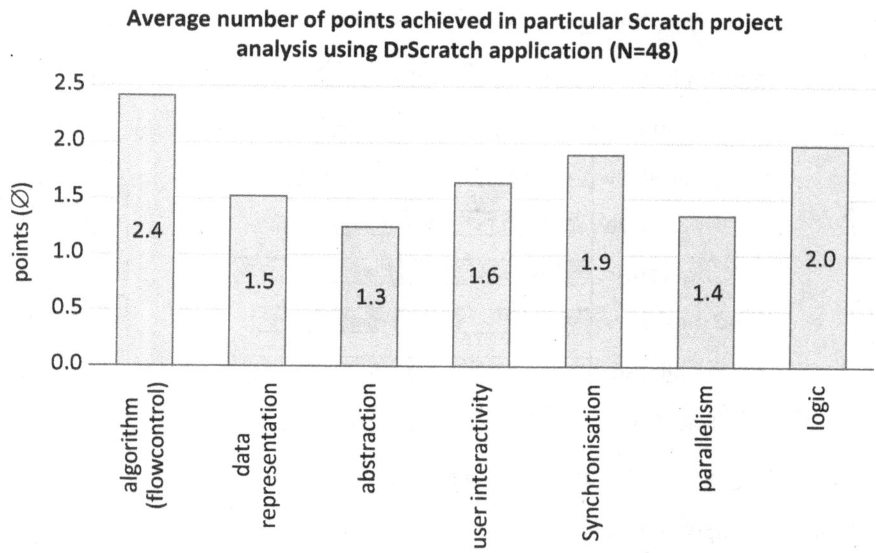

Figure 3.4 Average number of points given to Scratch projects to particular items in DrScratch.

Source: M. Svoboda (2018).

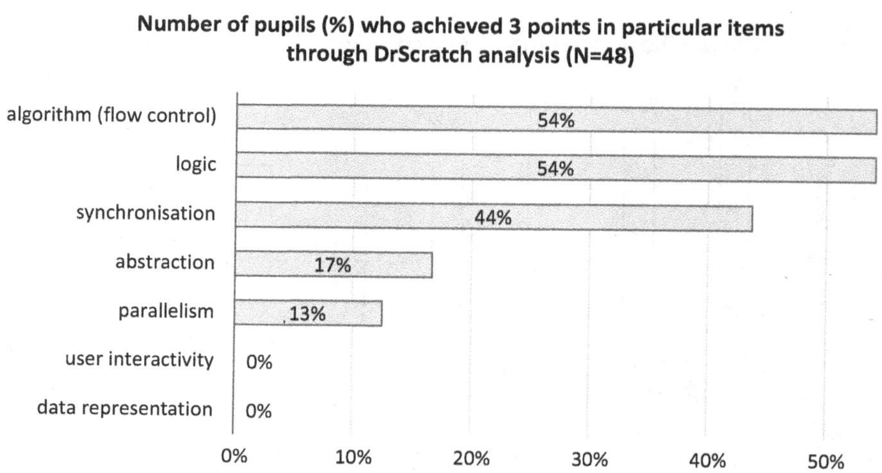

Figure 3.5 Items in which pupils received maximum points (3 points) in DrScratch analysis.

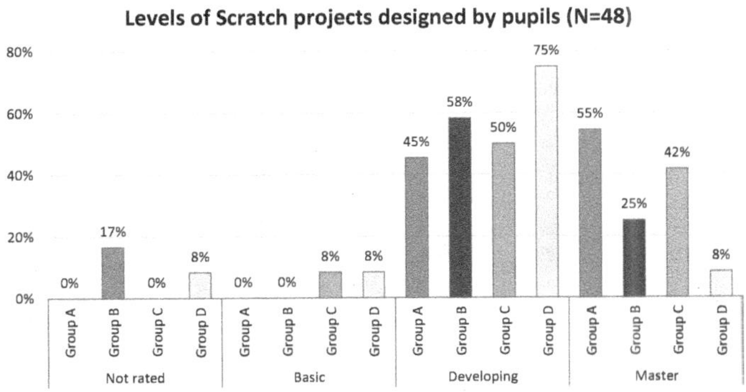

Figure 3.6 Levels of Scratch pupil's projects.
Source: M. Svoboda (2018).

Part 3: Results of tests A1 and A2

The degree of pupil's algorithmic thinking development has been assessed using data from the initial test A1 and the final test A2. The test A1 was identical to A2. The time difference between input and output testing was seven months.

Both tests A1 and A2 consist of 16 tasks. In 56% of test tasks, pupils demonstrate their *understanding of algorithm:* 56% of test tasks show how much pupils are able to *design an algorithm*. The tasks for tests A1 and A2 were designed by the authors, inspired by the iBeaver tasks (ibobr.cz).

In six out of nine tasks dedicated to understanding the algorithm, pupils achieved in test A2 better results; in three tasks (*Coloration of robots*, *Encrypt text*, and *Path*), however, their results in the test A2 were even worse than in A1 (Figure 3.7).

In seven out of the eight tasks dedicated to algorithm design, pupils achieved better results in test A2; in one task (*Coloration of robots*), however, their results in the test A2 were even worse than in A1 (Figure 3.8).

In solving problems focused on algorithmic tasks, 71% of pupils significantly improved their scores (measured by the number of points), 19% achieved the same number of points in test A1 and A2, and only 10% of pupils had their score worsen in test A2 than in test A1 (Table 3.4). After finishing the Scratch programming training, 29% of pupils did not show any demonstrable progress in solving problems of an algorithmic nature. Group A (Grade 5) was the most successful and was followed by group B (Grade 5). Comparing the results of each individual pupil showed that the best results were achieved mainly by pupils who scored a maximum of 10 points in the A1 test.

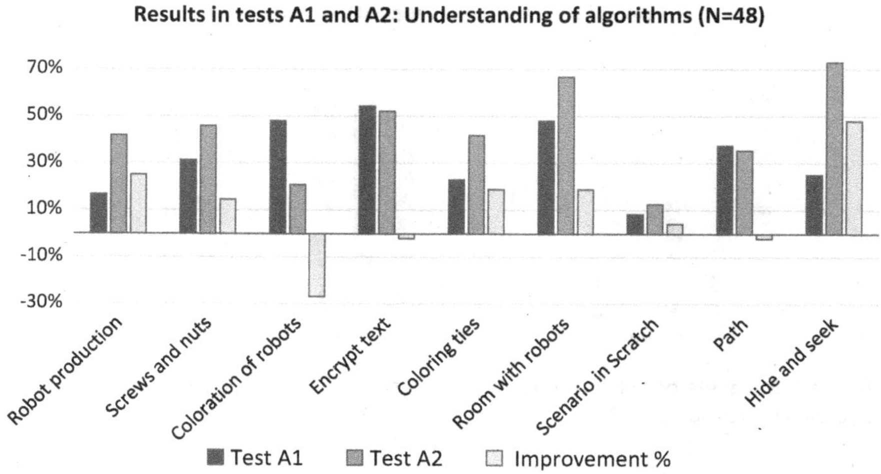

Figure 3.7 Comparison of results in tests AI and A2 by nine particular tasks focused on the understanding of algorithm.

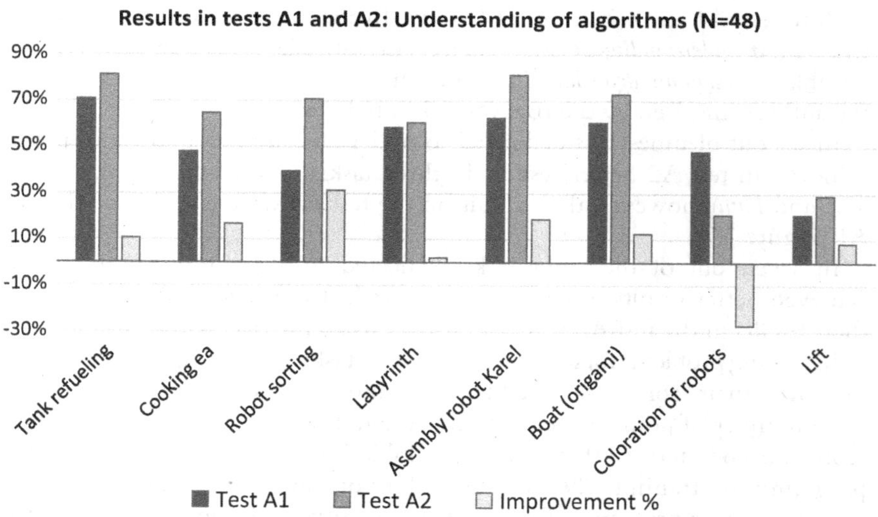

Figure 3.8 Comparison of results in tests AI and A2 by four particular tasks focused on algorithm design.

Table 3.4 The pupils' results in tests A1 and A2

Grade	*Group*	*Average score*		Δ	*Number of pupils who*		
		Test A1	*Test A2*		*Improved*	*Achieved the same result*	*Had underperformed*
Grade 5	Group A	7.5	11.0	+3.5	8	2	1
	Group B	6.5	8.7	+2.2	8	3	1
Grade 6	Group C	7.9	10.0	+2.1	10	1	2
	Group D	5.7	7.8	+2.1	8	3	1

Among pupils who were placed in the A1 test between the top nine, one pupil had the same number of points in test A2 as in test A1, and four pupils even did worse in the A2 test.

The *F*-test showed that the variance of the data sets in tests A1 and A2 did not differ significantly from each other. The *P*-value of the *T*-test is 0.001378 (< 0.05), which means that test A1 and test A2 differ statistically significantly.

Part 4: Results of questionnaire 2

Six weeks after the pedagogical experiment the pupils answered questionnaire 2 in which they evaluated their approach to programming and what they have learned with Scratch. About 70.9% of pupils enjoyed programming, 41.7% of pupils would like to continue in programming in Scratch, although the teaching of Scratch was over, a few of them (25%) would like to become a programmer (Figures 3.9 and 3.10).

Some pupils did not enjoy programming, especially some pupils of Grade 6.

> I did not enjoy the teaching of programming because I could not do it at all. Who should understand these commands?! There should be other commands. For example, 'Go to the top right corner' or 'Jump'. I liked the graphics, it was quite funny. And those unplugged algorithm activities; it was fun too.

Working with Scratch has influenced the way pupils learn (Figure 3.11). Some pupils felt they were actually playing rather than learning. Pupils preferred to work individually, not as a team; nevertheless, they enjoyed the peer interaction with their classmates. Pupils of Grade 5 rated teaching much more positively than pupils of Grade 6.

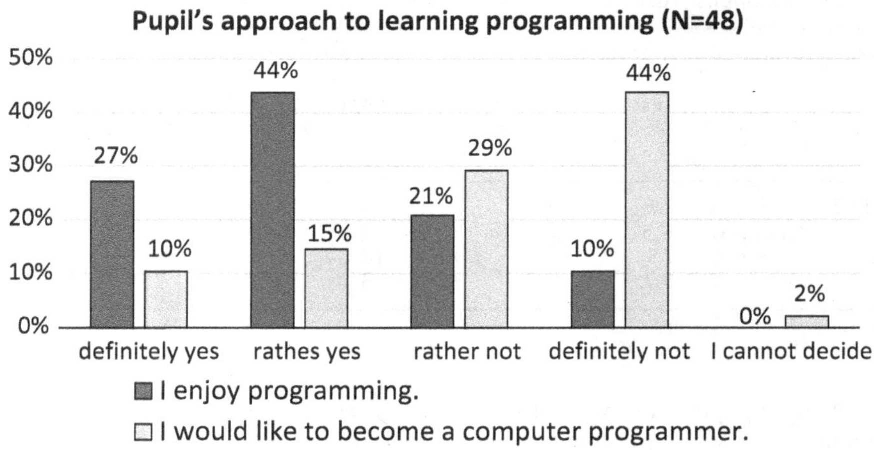

Figure 3.9 Pupil's approach to learning programming.
Source: M. Svoboda (2018).

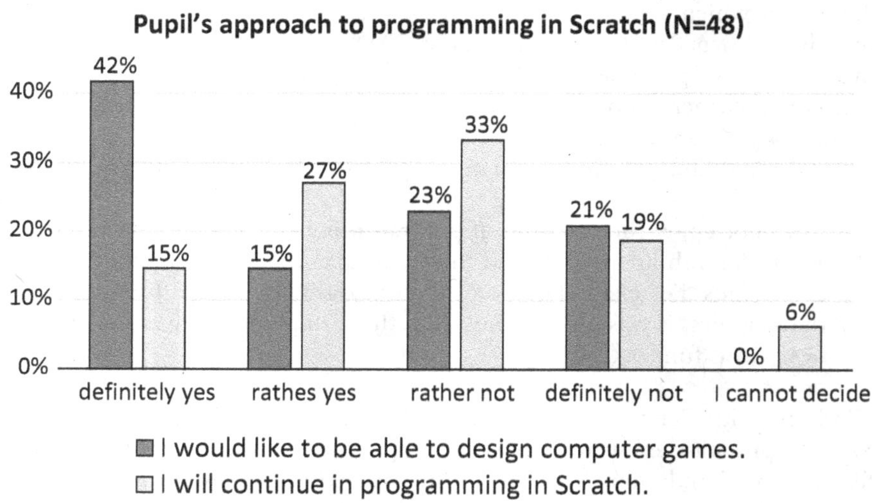

Figure 3.10 Pupil's approach to programming in Scratch.
Source: M. Svoboda (2018).

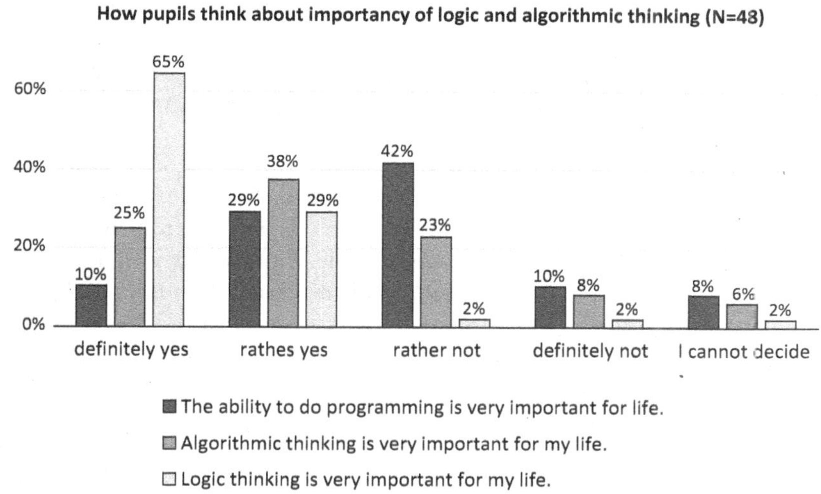

Figure 3.11 Importance of logic and algorithmic thinking from the pupils' perspective.
Source: M. Svoboda (2018).

Discussion

The major finding of this study is that although none of the pupils had any experience with Scratch 2.0, pupils' learning programming in Scratch achieved better results in the final test A2 than in A1 at the beginning. Comparing the results in both tests A2 and A1 focused on algorithmic tasks revealed that *at the end of the pedagogical experiment pupils improved in AT* in the final test A2; 34 from 48 pupils (70.8%) scored better in the test A2 than in A1. For the majority of test tasks in A2, the number of pupils who solved these tasks successfully was higher than in the initial test A1. There were only three test tasks (*Coloration of robots*, *Encrypt text*, and *Path*) for which the number of pupils who successfully completed these test tasks was lower in the test A2 than in the initial test A1. There could be several explanations: Firstly, these tasks could be very difficult for pupils; secondly, **the strategy used in these test tasks differ from the strategy used in Scratch programming**, and consequently, programming may not have any influence on solving these three tasks in A2.

Another important finding is that pupils enjoyed working in Scratch and that they consider logical thinking the ability to design algorithms and programming important for their lives: 70.8% pupils enjoyed programming in Scratch so much that a quarter of them would like to become programmers and 12.5% pupils continued programming even after the pedagogical experiment was over; 70.8% pupils stated they understand coding; 48.0% pupils enjoyed designing algorithms; 45.8% pupils would like to learn more about programming in Scratch; 62.5% pupils think that AT plays an important role for their life; and 68.7% pupils are convinced that informatics is important for their future adult professional life.

Programming in Scratch, especially programming one's own projects, induces learning that involves complex thinking processes, creative, and learning activities that foster "skills such as problem-solving, confronting 'failures', and strategies to explore and to decide possible solutions, as well as to structure thoughts and actions" (Bers et al., 2014 in Papavlasopoulou et al., 2018, p. 51). Programming in Scratch contributes to promoting *active learning.* In this study, 81.2% of pupils consider the knowledge and skills they have acquired as important. The final games programmed in Scratch 2.0 fostered pupils' analytical thinking and the skills needed for problem-solving.

During programming, the pupils learned to detect errors, to think about how to "describe" a process to solve a problem, to identify mistakes in their own thinking, and to explain and interpret algorithms. All of these experiences pupils can apply as a learning strategy and strategy for solving (similar) problems at any time in the future. Algorithm design has led pupils to be able to think in a structured and exact way and to express outcomes using symbolics. It demonstrates that thinking can be made visible. The ability to find patterns in structures and cycles in processes can be exploited in other pupil's activities and learning. AT required pupils' logical and precise thinking and improved their ability to decompose a problem into small parts, to apply pattern recognition, and to develop step-by-step instructions for solving the problem. All these skills presuppose deep thinking processes.

That some pupils enjoyed programming of their computer games in Scratch so much that they continued in programming also outside school may be considered a great benefit from the pedagogical experiment; the worlds of formal and informal learning were connected.

Teaching Grade 5 and 6 pupils has shown that Seymour Paper's idea that a procedure in a programming environment (Scratch in this case) "can become something named, manipulated and recognized; terming it 'an object to think with' has been fulfilled" (Kalaš & Benton, 2017, p. 568).

Benton et al. (2018) focused on examining if programming benefits pupil's mathematical understanding and on how pupils understand the concept of algorithm and how they imagine it. In this study, pupils should also show what they understand by an algorithm. In the introductory lesson, they were asked to describe the process(es) they know from everyday life as the making of a cup of tea, or cooking scrambled eggs, frying pancakes, or light bulb change. Later, the pupils analysed descriptions of these processes and whether their instructions could be repeated and whether the processes were complete.

Analysis of pupil projects and test B1 have shown that, in terms of Bloom's taxonomy, it has been successful, like Meerbaum-Salant et al. (2013, p. 245) "in meeting the learning objectives, especially in the area of applying (the ability to execute programs or algorithms, to track them, and to recognize their goals) and creating (the ability to plan and produce programs or algorithms)".

Tests A1 and A2 were used as research tools to compare pupils' level of algorithmic thinking at the beginning and at the end of learning programming. The tests were pilot-tested with children at a summer camp; the tests cannot, however, be considered as standardized tests. The pilot-testing highlighted some difficulties and formulation ambiguities which brought a number of suggestions for their final editing.

Some problems (often due to the misunderstanding of some mathematical concepts) have been encountered during the pedagogical experiment including pupils' complex reflections on the algorithmization of the tasks and subsequent coding. Pupils were faced with problems in determining the position of sprites using a coordinate system in Scratch. This problem has been reflected in the solution of the Task 7 in test B1. In this task, pupils should select a sequence of commands to move a sprite from the centre of the scene down to the left (with x = −100, y = −100). The problem could be that the pupils did not know the concept of a negative number. The description of a sprite's position using the x and y coordinates was also incomprehensible for some pupils. Another relatively fundamental problem was ignorance of the division of a circle into angles and how to express a geometric angle in degrees.

"Computing education is a newcomer to the rank of discipline-based education research" (Cooper et al., 2014, p. 35). "We know relatively little about foundation of computing knowledge" (Cooper et al., 2014, p. 36). Denning (2017, p. 38) recommends "to teachers and education researchers: Use Aho's historically well-grounded definition and use competency-based skill assessments to measure student progress". To obtain more accurate data about pupil's algorithmic thinking, tests A1 and A2 would require standardized tests.

Conclusion

The study showed that programming in Scratch has a positive influence and impact on pupil's algorithmic thinking. Therefore, it is important to implement educational programming into school curricula and to concentrate, as Benton et al. (2018, p. 591) recommend, on the following

> stages of learning in AT: interpret, step through, and predict the outcome of a given algorithm (understanding); generate own algorithm to achieve a desired result (design); and adapt an algorithm for solving a specific problem to more general problems (generalization / simplification).

References

Aho, A. V. (2012). Computation and computational thinking. *The Computer Journal, 55*(7), 832–835.

Benton, L., Kalaš, I., Saunders, P., Hoyles, C., & Noss, R. (2018). Beyond jam sandwiches and cups of tea: An exploration of primary pupils' algorithm-evaluation strategies. *Journal of Computer Assisted Learning, 34*, 590–601.

Bers, M. U., Flannery, L., Kazakoff, E. R., & Sullivan, A. (2014). Computational thinking and tinkering: Exploration of an early childhood robotics curriculum. *Computers & Education, 72*, 145–157.

Cooper, S., Grover, S., Guzdial, M., & Simon, B. (2014). A future for computing education research: Seeking to address the most important issues facing the computer education research community. *Communications of the ACM, 57*(11), 34–36.

Creswell, J. W. (2012). *Educational research: Planning, conducting, and evaluating quantitative and qualitative research* (4th ed.). Boston, MA: Pearson Education.

Denning, P. J. (2017, June). Remaining trouble spots with computational thinking. Addressing unresolved questions concerning computational thinking. *Communications of the ACM, 60*(6), 33–39.

Dewey, J. (2017). Coding in the curriculum. Salem Press Encyclopedia [online]. 2017 [cit. 2019-03-01].

Durak, Y. H. (2018). The effects of using different tools in programming teaching of secondary school students on engagement, computational thinking and reflective thinking skills for problem solving. *Technology, Knowledge and Learning*, 1–17.

Fellows, M., Bell, T., & Witten, I. (2005). Computer science unplugged. Retrieved from https://csunplugged.org/en

Futschek, G. (2006). *Algorithmic thinking: The key for understanding computer science.* In Lecture Notes in Computer Science 4226, Springer, pp. 159–168.

Grover, S., & Pea, R. (2013). Computational thinking in k–12: A review of the state of the field. *Educational Researcher, 42*(1), 38–43.

Grover, S., & Pea, R. (2018). Computational thinking: A competency whose time has come. In S. Sentence, E. Barendsen & C. Schulte (Eds.), *Computer science education. Perspective on teaching and learning in school* (pp. 19–38). London: Bloomsbury Academic.

Guzdial, M. (2008). Education: Paving the way for computational thinking. *Communications of the ACM, 51*(8), 25–27.

Hromkovič, J., Kohn, S., Komm, D., & Serafini, G. (2017). Algorithmic thinking from the start. In: Bulletin of the EATCS 121, The Education Column.

Hromkovič, J., & Lacher, R. (2017). How to convince teachers to teach computer science even if informatics was never a part of their own studies. *Bulletin of the EATCS*, 123.

Hsu, T-C., Chang, S-C., & Hung, Y-T. (2018). How to learn and how to teach computational thinking: Suggestions based on a review of the literature. *Computers & Education, 126*, 296–310. doi:10.1016/j.compedu.2018.07.004

Kalaš, I., & Benton, L. (2017). Defining procedures in early computing education. In A. Tatnall & M. Webb (Eds.), *Tomorrow's learning: Involving everyone. Learning with and about technologies and computing* (pp. 567–578). Cham, Switzerland: Springer.

Lahtinen, E., Ala-Mutka, K., & Järvinen, H. M. (2005). A study of the difficulties of novice programmers. *ACM SIGCSE Bulletin, 37*(3), 14–18.

Maloney, J., Resnick, M., Rusk, N., Silverman, B., & Eastmond, E. (2010). The Scratch programming language and environment. *ACM Transactions on Computing Education* (TOCE), *10*(4), 16.

Meerbaum-Salant, O., Armoni, M., & Ben-Ari, M. (2013). Learning computer science concepts with Scratch. *Computer Science Education, 23*(3), 239–264.

Moreno-León, J., Robles, G., & Román-González, M. (2015). Dr. Scratch: Automatic analysis of Scratch projects to assess and foster computational thinking. *Revista de Educación a Distancia, 15*(46), 1–23.

Papavlasopoulou, S., Sharma, K., & Giannakos, M. N. (2018). How do you feel about learning to code? Investigating the effect of children's attitudes towards coding using eye-tracking. *International Journal of Child-Computer Interaction, 17*, 50–60.

Papert, S. (1980). *Mindstorms—Children, computers and powerful ideas*. New York: Basic Books, Inc.

Robins, A., Rountree, J., & Rountree, N. (2003). Learning and teaching programming: A review and discussion. *Computer Science Education, 13*(2), 137–172.

Rogalski, J., & Samurcay, R. (1990). Acquisition of programming knowledge and skills. In J.-M. Hoc, T. R. G. Green, R. Samurcay, & D. Gilmore (Eds.), *Psychology of programming* (pp. 157–174). London: Academic Press.

Svoboda, M. (2018). *Rozvoj algoritmického myšlení žáků ZŠ ve výuce informaticky zaměřených předmětů s využitím Scratch.* (The development of algorithmic thinking in elementary school pupils using the Scratch programming tool.) Univerzita Karlova, Pedagogická fakulta. Praha, 2018. Diplomová práce. Retrieved from https://is.cuni.cz/webapps/zzp/download/120307515

Tekerek, M., & Altan, T. (2014). The effect of Scratch environment on student's achievement in teaching algorithm. *World Journal on Educational Technology, 6*(2), 132–138.

Weintrop, D., Behesht, E., Horn, M., Orton, K., Jona, K., Trouille, L., & Wilensky, U. (2016, February 1). Defining computational thinking for mathematics and science classrooms. *Journal of Science Education and Technology, 25*(1), 127–147. doi:10.1007/s10956-015-9581-5

Weintrop, D., & Wilensky, U. (2017). Comparing block-based and text-based programming in high school computer science classrooms. *ACM Transactions on Computing Education, 18*(1), 3:1–3:25.

Wing, J. M. (2006, March). Computational thinking. *CACM Viewpoint*, 49(3), 33–35.

Chapter 4

Digital educational games in primary education

Revisiting the results of the research projects of the ETiE initiative

Emmanuel Fokides

Introduction

The research initiative Emerging Technologies in Education (ETiE) was launched in 2015. Although not a formal project, a substantial number of studies were conducted under its umbrella. ETiE's main objective is to examine the results of the educational uses of emerging technologies in primary and high school settings. In an educational context, emerging are technologies that recently found their way to classrooms and their instructional value is yet to be unveiled. For example, ETiE examined the educational uses of drones, virtual reality, 3D multi-user virtual environments, and augmented reality. Also, a portion of the research effort focused on digital games, because it was considered that their use in education presents interesting and still unexplored aspects.

Young people spend a substantial part of their leisure time by playing digital games, which are the predominant entertainment medium even of very young children (OfCom, 2013). At the same time, it is generally accepted that digital games can play a notable role in education. Indeed, digital game-based learning (Prensky, 2001) can be applied in all levels of education and in almost all courses (Nie et al., 2014). The relevant literature reports learning gains and improved conceptual understanding, increased motivation for learning (Ke, 2008), development of a number of skills, and an impact on creativity (Hsiao et al., 2014).

Given the above, it was considered interesting to examine under which conditions the strong relationship between children and digital games can be exploited to promote the learning objectives related to the teaching of various subjects in primary school. As a result, a series of research projects were designed and implemented over the past couple of years, involving the use of digital games to examine whether their use in teaching results in better learning outcomes compared to other teaching tools. Having accumulated a number of such projects, 11 in total, this chapter revisits their results and critically re-evaluates them.

Digital games in education

Students are increasingly dependent on technology for information retrieval/exchange and for communicating (Spires, 2008). Reasonably enough, appealing digital environments, such as digital educational games (DEGs), are compatible with their skills and interests. A multitude of definitions has been given for DEGs, having in common the idea that they are games that offer entertainment but, at the same time, have an instructional value (Susi et al., 2007).

There is a consensus in the literature that digital games can become vital teaching/learning tools (Guillén-Nieto & Aleson-Carbonell, 2012). Indeed, Prensky (2001) supported the view that there is actually no distinction between games and education. Their use in education is embraced by almost all learning theories (Braghirolli et al., 2016). On one hand, many games seek to train students in concepts or skills using repetitive practices, thus, realizing behavioural principles. In this case, repetition reinforces responses and increases the likelihood of another occurrence of the desired behaviour when the stimulus is present again. On the other hand, DEGs based on constructivist perceptions, seek the active participation of the player/student in the learning process, so that the new knowledge is constructed through the game (Shute et al., 2011). In this case, the purpose of using DEGs is to achieve a student-centred interactive experience that changes the relationship between the student, the media, and the teacher. Finally, Game-Based Learning seeks to balance the learning subject, the game, and the players' ability to maintain and apply what they have learned. This concept transcends the development of games just for students to play, as it is about designing learning activities that gradually introduce concepts and guide users towards a final goal (Pho & Dinscore, 2015).

There are many examples about the pedagogical utilization of digital games and the results they yielded, covering a wide range of learning subjects such as sciences, mathematics, and language (e.g., Bakker et al., 2014; Hummel et al., 2010). Most of them converge on the fact that digital games have a lot to offer if they are systematically exploited and when the objectives of the game are directly related to the learning goals (Sung & Hwang, 2013). In addition to the satisfactory results in knowledge acquisition, the following have also been reported: An increased effort to achieve the learning goals, active participation, enjoyment, reinforcement of motivation for learning, and improvement of students' attitudes to either specific subjects or to education as a whole (e.g., Ke, 2008; Robertson, & Miller, 2009; Tüzün et al., 2009). The development of collaborative capacities, the reinforcement of problem-solving skills, as well as the development of specialized knowledge and skills (Connolly et al., 2012) have also been mentioned.

Instead of using ready-made digital games, a body of research examined how game authoring enabled students to understand programming concepts and practices. Game authoring has been mainly studied in tertiary and secondary education (e.g., Kazimoglu et al., 2012), but the literature is not extended. Studies in which the target group was primary school students are scarce (e.g., Baytak & Land, 2010) and the same applies to studies that examined the effects of computer game development as a pedagogical activity (Owston et al., 2009).

There are a number of explanations for the effectiveness of DEGs. For example, it has been observed that when students play educational games, they tend to spend more time in trying to learn, and, in turn, this affects learning outcomes (Tobias et al., 2011). Also, students pay more attention to a learning activity when it is presented through a game (Garris et al., 2002). Another element is the direct feedback they offer; students can immediately see the results of their actions (e.g., if they answered a question correctly). Thus, students are encouraged to explore, experiment, and discover new concepts and strategies (Kirriemuir, 2002).

However, the introduction of DEGs in education is not free of problems. While in many studies improved learning outcomes were reported, there are studies in which the outcomes were neutral or even negative (e.g., Perrotta et al., 2013). Also, researchers pointed out that the pedagogy of DEGs is not well-developed (Ulicsak & Williamson, 2011) and that learning with games has to be supported by effective instructional measures (Egenfeldt-Nielsen, 2006). Finally, perhaps the most important problem is the effort, time, and cost required for developing DEGs (Westera et al., 2008).

A brief presentation of the projects and of their results

As already mentioned, from 2015 to 2018, as part of the ETiE initiative, 11 research projects related to the use of DEGs were carried out, having as a target group primary school students. Their main objective was to examine whether the use of DEGs enables students to achieve better learning outcomes compared to other teaching tools, as suggested by the literature. In five projects, programming concepts were taught to students aged 7–9 (135 individuals), 8–9 (75 individuals), 10–11 (138 individuals), and 11–12 (130 individuals). In four projects maths was the subject and the target groups were students aged 6–7 (129 individuals), 9–10 (189 individuals), and 11–12 (66 individuals). One project had as a theme the teaching of English as a foreign language (60 students, 10–11 years old) and, in another, modules from the study of an environment course were taught (54 students, 8–9 years old).

A quasi-experimental design was applied to all projects since data from whole classes were collected. Also, in all cases, the participating students were split into three groups (one control and two experimental). In all but five projects (in which programming was taught), the teaching in the control groups was conventional, using a teacher-centred method and the school textbooks, thus, reflecting the way students are usually taught, on a daily basis, in a Greek primary school. In the experimental groups, combinations of various instructional methods were applied, that were based on the principles of constructivism (e.g., group work and active participation of students). Such methods were (a) without the use of games, (b) exclusively with the use of games, without the teachers' active participation, and (c) with games and with the teachers' active participation. In all the above interventions and in all groups, each session lasted for two teaching hours (one and a half hour in total). The number of sessions varied from three to six (in each group), depending on the learning subject and the availability of teaching hours. A summary of the projects can be found in the Appendix, while the complete data analysis of all projects can be found at http://opensimserver.aegean.gr/data for publications/Data games_EN.htm.

It should be noted that in all cases Microsoft's Kodu Game Lab (www.kodugamelab.com/) was used for the games' development. Kodu allows the rapid and relatively easy development of 3D cartoonish games, through visual programming using a set of very simple rules that are based on natural terms and concepts such as "see", "hear", and "bump". It is also very important to stress that the games were not developed by groups of experts, but by the teachers who participated in the projects. For that matter, they attended seminars and the research team provided technical assistance and guidance, but there was no further involvement in the type and philosophy of the games the teachers developed. The reason for this approach was to check whether teachers are capable of being producers rather than consumers of educational software.

Coming to the projects in which programming was taught, Kodu was used once again. In three of them (see Appendix, No. 7, 8, and 9) specific programming concepts were taught (e.g., sequences, variables, loops, and subroutines). Two were addressed to very young students and, in these projects, one of the control groups used a board game instead of printed material. In the last two cases (No. 10 and 11), students were asked to develop their own games and they were taught all the available programming concepts as well as game design. Moreover, the projects' duration was large (35 and 50 two-hour sessions, respectively). While the sample was once again divided into three groups, this time what students can achieve was checked either by using only notes, with a limited, or an active teachers' participation in the process.

In all projects, with the exception of the last two related to programming, evaluation sheets were used for collecting data and their number was equal to the number of sessions. Pre-tests were used, for establishing students' initial knowledge level (allowing a better interpretation of results). Also, delayed post-tests, administered two weeks after the end of each project, established the sustainability of knowledge. The evaluation sheets included questions of escalating difficulty, examining not only knowledge acquisition, but also whether students were able to apply the new knowledge in other situations, to associate different pieces of information/knowledge, and make informed decisions. In the last two projects, the students' games were evaluated by analysing their content both in terms of programming and game design adequacy. Finally, in most projects, short questionnaires were administered to the groups of students that used or developed games, for recording their views and attitudes towards the use of games in their teaching.

As for the projects' results, these were interesting, to say the least. A key conclusion is that in all but one project, the use of DEGs (with or without the active participation of teachers) resulted in better learning outcomes than conventional teaching. In projects with English as a foreign language as the subject, the use of DEGs without the active participation of the teachers had equally good results with well-organized constructivist teaching without the use of games. In the project where units from the study of the environment were taught, the groups that used games (one with the active participation of the teacher and one without) had equally good results and the results of both were better than conventional teaching. In one out of the four projects where maths was taught, DEGs and conventional teaching had equally good results; in the other three, DEGs surpassed conventional teaching. Then again, well-organized constructivist teaching methods and DEGs had equally good results.

As far as the programming projects were concerned, the development of digital games led to a better understanding of programming concepts compared to both conventional and constructivist teaching. However, the most interesting results were obtained from the two interventions that had a long duration (70 and 100 hours). In both cases, it seems that while the teaching method initially played a role, over time even the students who had only notes and received no other support, managed to develop equally good games and with equally few mistakes with the other groups that received partial or significant support by their teachers.

Finally, on the basis of the results in the questionnaires for recording students' views and attitudes, it was clear that, in all cases, the participating students found that teaching with games was interesting and fun. In addition, they were more motivated to learn, no problems related to the use of games were reported, and collaboration seemed to have worked well.

Discussion

The projects' results are in line with the existing literature, which pointed out that the use of DEGs results in the same or better learning gains compared to other conventional forms of instruction (e.g., Fokides, 2018; Sung & Hwang, 2013). A number of reasons, related to the games per se and to the teaching methods that were followed, may have contributed to these results. The first fact is that students, even the younger ones, did not experience any problems when playing the games, confirming the special relationship children have with technology (Prensky, 2001). On the contrary, students considered the whole process amusing, a fact that emerged in the analysis of all projects' questionnaires. The fun and enjoyment may be attributed to Kodu's graphics and cartoonish characteristics, which appear to attract the interest of younger children (Fowler, 2012). The link between the playful character of the teaching because of the use of DEGs and the improved conceptual understanding has been highlighted by others (e.g., Mawer & Stanley, 2011). Moreover, the increased motivation for learning when playing DEGs (Ke, 2008), was also confirmed by students' responses to the relevant questions. Students' interest may have been intensified by the fact that the games included scoring systems, which "rewarded" them upon their successful completion of a task. Furthermore, the scores provided students' immediate feedback for the results of their actions. The control of the learning process, through constant feedback, which DEGs allow, has been identified by previous research (McClarty et al., 2012).

Working in pairs and students' collaboration formed the basis of all projects. Students' responses to the relevant questions indicated that they highly appreciated their collaboration with their peers. Therefore, it seems that DEGs offered a fertile ground for the development of collaborative activities (Sauvé et al., 2010) and together with active engagement, and experimentation, students were able to attain good learning results (Westera et al., 2008). In a number of projects (i.e., in projects 1, 2, 3, and 6), the teachers' role was purposefully minimized, allowing increased students' autonomy and control over their learning process. Given that even in these projects the learning outcomes were better than conventional teaching, confirmed the views of those who (a) considered autonomy and collaboration as factors that work alongside when playing DEGs (Fokides, 2018) and (b) asserted that when students have a high degree of autonomy, positive learning outcomes are to be expected (Nunes et al., 2009). Moreover, the learning outcomes in these groups and in groups in which students had the help and guidance of their teachers (with or without the use of games) were equally good. This finding may lead to the assumption that games were so effective that they successfully assumed the teachers' role, demonstrating the power of Game-Based Learning (Sung & Hwang, 2013).

The results in the projects related to programming (projects 7–11) are similar to the findings of previous research, which underlined the contribution of Kodu in making the learning process more enjoyable and in helping students to have a better understanding of the basics of programming (e.g., Shokouhi et al., 2013). It seems that the effectiveness of Kodu is mainly due to its playful character; after all, its main purpose is to develop games. Also, students did not encounter any particular problems. On the basis of the above, it can be argued that Kodu is quite an effective tool compared to conventional means.

The results in projects concerning programming that also had an extended duration (projects 10 and 11) were challenging, as they contradicted the majority of the existing literature. Indeed, it is suggested that when learning a programming language, direct instruction, formal introduction, and demonstration of programming concepts are needed (Denner et al., 2012). There is also the view that game authoring does not allow the understanding of more complex concepts, at least without explicit teaching (Denner et al., 2012). Contrary to the above, it was found that even without systematic instruction, students learned and used quite a lot of programming concepts and successfully developed functional digital games. The critical factors in both projects were time and collaboration. Thus, the views of Kafai (2012), who supported the implementation of extended projects for teaching programming, were verified by the findings of these two studies.

Implications for practice

The above results have several implications for educators as well as for education policy-makers. First, it should be reminded that the games were developed by the teachers who participated in the projects. In Greece, teachers who wish to introduce innovative teaching methods with the use of technological tools are left without substantial support and this holds true for DEGs as well. Thus, given the lack of DEGs that have been certified for their educational value, the burden of developing them is passed on to the teachers. Indeed, some very interesting observations emerged from this effort. First, the time required for the games' development ranged from 50 to 150 hours (depending on the number of sessions and the games' complexity), which, in any case, is considerable. At the same time, the teachers encountered difficulties in materializing their ideas, due to Kodu's limited number of available objects/characters. This led to the need to devise other ways of presenting the cognitive material, which required reflection and experimentation. Therefore, the effort required for the development of DEGs by "amateurs" is disproportional in relation to the final product (Kluge & Riley, 2008).

In addition, all games were, in essence, "trial and error" applications, realizing behaviorism's concepts, although the teaching framework was based on constructivism. It is also true that an expert could characterize the games as incomplete or that they did not implement the learning objectives correctly, which may have adversely affected the learning outcomes. On the other hand, since "amateurish" DEGs had quite good results, one might assume that DEGs developed by experts might have been able to produce even better results. In this respect, the need for collaboration between educators and ICT specialists for the development of DEGs is emphasized. That is because the former have the necessary pedagogical knowledge, but not the necessary ICT skills, while the latter have the appropriate technical background, but are lagging behind in understanding the pedagogical principles. Moreover, if we want teachers to be able to develop their own DEGs, it is necessary to make available to them tools that make the whole process much more flexible and accessible to the average user (Scacchi, 2012).

With regard to the projects relating to programming, interesting implications can be noted. A major problem was the large number of sessions that caused considerable disruption in the schools' timetables and there were justified complaints by the teachers. Thus, it is necessary to find a balance between the duration and the need to apply an effective teaching strategy, such as the one implemented in the projects (through the development of digital games by the students) (Ke, 2014). Of course, to achieve this, it is necessary for the educational policy-makers to revise the primary school's curriculum and increase the teaching hours dedicated to ICT-related subjects.

Another issue to consider is the suitability of Kodu for teaching programming concepts to primary school students. Although some programming concepts are easily implemented and the students found it amusing, the programming language is very different from the usual ones. Thus, some argued that Kodu is more appropriate for instilling the initial interest of students for programming or for teaching game design principles (Morris et al., 2017). On the other hand, Scratch, which is widely used for the systematic teaching of programming, has also been criticized (Meerbaum-Salant et al., 2013). The issue is not an easy one to resolve and largely depends on the objectives set.

The last issue to be discussed is who can teach programming at the primary level. In all five projects, this task was undertaken by the classes' teachers and not by the ICT teachers. There were specific necessities that imposed this arrangement. ICT teachers share their working hours in many classes if not in different schools. So, it is impossible for them to devote several consecutive hours to a class, as the projects required. On the other hand, it appeared that the teachers, with a relatively short training, could carry out this task. Therefore, this is also a matter for the education policy-makers to consider.

Conclusion

The results of the studies presented in the previous sections were at least interesting. On the other hand, there are limitations to the generalizability of the results. The sample sizes, on several occasions, although adequate for statistical analysis, could have been larger. Also, all projects were conducted in Greek primary schools. The teaching subjects were limited, mainly, to maths and programming. Moreover, in nine cases, data were collected using quantitative tools (questionnaires and evaluation sheets). The above restrictions set the framework for future studies. Larger sample sizes, a wider variety of teaching subjects, and different age groups can provide useful information on the impact of digital games. Quantitative as well as qualitative data collection tools, such as interviews and observations, will allow researchers to better understand the impact of digital games on education. Finally, comparisons between digital games and other technological tools can provide a better picture of their relative advantages (or disadvantages). ETiE is already planning the next wave of research interventions following the above guidelines.

In conclusion, DEGs are indeed an interesting alternative method for teaching several subjects to primary school students. However, there is still a long way ahead until their education potential is fully realized.

Appendix: The projects in brief

No	*Subject*	*Target group (ages)*	*Sample size*	*Groups*	*Duration (two-hour sessions)*	*Results*
1	English as a foreign language	10–11	60	1 Conventional teaching 2 Group work, constructivist based, active participation of teachers and students 3 Games, group work, constructivist based, active participation of students, no teachers' participation	5	Findings: Group 3 outperformed Group1 in zero cases and Group 2 in three cases. In all the other cases, as well as in the delayed post-test, there were no statistically significant differences. Interpretation: Games produced better results compared to constructivist teaching without games and equally good compared to conventional teaching. Positive attitudes of students towards games.
2	Science	8–9	54	1 Conventional teaching 2 Games, group work, constructivist based, active participation of teachers and students 3 Games, group work, constructivist based, active participation of students, no teachers' participation	3	Findings: Group 3 outperformed Group 1 in all cases and Group 2 in zero cases. Group 2 outperformed Group 1 in all cases. In the delayed post-test Group 3 outperformed Group 1 and Group 2. Interpretation: Games produced better results compared to conventional teaching even without the active participation of teachers. Given that Groups 2 and 3 had equally good results, the teacher had no impact on students' learning outcomes.
3	Maths	9–10	63	1 Conventional teaching 2 No games, group work, constructivist based, active participation of teachers and students 3 Games, group work, constructivist based, active participation of students, no teachers' participation	4	Findings: Group 3 outperformed Group 1 in two cases and Group 2 in zero cases. In the delayed post-test Group 3 outperformed Group 1. Interpretation: Games produced better results compared to conventional teaching and equally good compared with constructivist teaching, even if in Group 3 the teachers had no active role. Positive attitudes of students towards games. Students were motivated to learn.

4	Maths	6–7 9–10 11–12	69 66 66	1 Conventional teaching 2 No games, group work, constructivist based, active participation of teachers and students 3 Games, group work, constructivist based, active participation of students and teachers	4 4 4	Findings: Group 3 outperformed Group 1 in 12 cases and Group 2 in 3 cases (and was outperformed by this group in 2 cases). In the delayed post-test Group 3 outperformed Group 1 in 3 cases and Group 2 in 1 case. Interpretation: Games produced better results compared to conventional teaching and equally good compared with constructivist teaching but without games. Given that in Groups 2 and 3 the teaching was constructivist based, the impact of the games was minimal. Positive attitudes of students towards games. Students were motivated to learn.
5	Maths	9–10	60	1 Conventional teaching 2 Games, group work, constructivist based, active participation of students, no teachers' participation 3 Games, group work, constructivist based, active participation of students and teachers	3	Findings: Group 3 outperformed Group1 in one case and Group 2 in one case. Group 2 outperformed Group 1 in zero cases. In the delayed post-test Group 3 outperformed Group 1. Interpretation: No clear advantage of games over conventional teaching. Positive attitudes of students towards games. Students were motivated to learn.
6	Maths	6–7	60	1 Conventional teaching 2 Board games, group work, constructivist based, active participation of students and teachers 3 Games, group work, constructivist based, active participation of students, no teachers' participation	6	Findings: Group 3 outperformed Group 1 in three cases and Group 2 in two cases. Group 2 outperformed Group 1 in zero cases. In the delayed post-test Group 3 outperformed Group 1 and Group 2. Interpretation: Games produced better results compared to conventional teaching and equally good compared to constructivist teaching with a board game, even if in Group 3 the teachers had no active role.

(Continued)

No	*Subject*	*Target group (ages)*	*Sample size*	*Groups*	*Duration (two-hour sessions)*	*Results*
7	Programming	8–9	75	1 Conventional teaching 2 Board game, group work, constructivist based, active participation of students and teachers 3 Games, group work, constructivist based, active participation of students and teachers	3	Findings: Group 3 outperformed Group 1 in three cases and Group 2 in two cases. In the delayed post-test Group 3 outperformed Group 1 and Group 2. Interpretation: Games produced better results compared to conventional teaching and also better compared to constructivist teaching with a board game.
8	Programming	7–9	135	1 Conventional teaching 2 Board game, group work, constructivist based, active participation of students and teachers 3 Games, group work, constructivist based, active participation of students and teachers	3	Findings: Group 3 outperformed Group 1 in six cases and Group 2 in four cases. In the delayed post-test Group 3 outperformed Group 1 and Group 2. Interpretation: Games produced better results compared to conventional teaching and also better compared to constructivist teaching with a board game.
9	Programming	11–12	60	Students were asked to develop games 1 Group work, no systematic teaching 2 Group work, conventional teaching 3 Group work, constructivist based, active participation of students and teachers	3	Findings: Group 3 outperformed Group 1 in two cases and Group 2 in three cases. Interpretation: Games produced better results compared to conventional teaching and also better compared to constructivist teaching with no use of games.

10	Programming	11–12	70	Students were asked to develop three games of escalating complexity 1 Group work, no notes, no systematic teaching 2 Group work, notes, no systematic teaching 3 Group work, notes, systematic teaching	35	Findings: Both Group 1 and Group 3 outperformed Group 2 in one case. Interpretation: No clear advantage of one method over the others. Positive attitudes of students towards games. Students were motivated to learn.
11	Programming	10–11	138	Students were asked to develop three games of escalating complexity 1 Group work, notes, no systematic teaching 2 Group work, notes, limited teachers' participation 3 Group work, notes, active teachers' participation	50	Findings: In Game 1, Group 3 outperformed the other groups. In Game 2, Group 3 outperformed only Group 1. In Game 3 there were no statistically significant differences. Interpretation: No clear advantage of one method over the others. Positive attitudes of students towards games. Students were motivated to learn.

References

Bakker, M., van den Heuvel-Panhuizen, M., & Robitzsch, A. (2014). First-graders' knowledge of multiplicative reasoning before formal instruction in this domain. *Contemporary Educational Psychology, 39*(1), 59–73.

Baytak, A., & Land, S. M. (2010). A case study of educational game design by kids and for kids. *Procedia-Social and Behavioral Sciences, 2*(2), 5242–5246.

Braghirolli, L. F., Ribeiro, J. L., Weise, A. D., & Pizzolato, M. (2016). Benefits of educational games as an introductory activity in industrial engineering education. *Computers in Human Behavior, 58*, 315–324.

Connolly, T. M., Boyle, E. A., MacArthur, E., Hainey, T., & Boyle, J. M. (2012). A systematic literature review of empirical evidence on computer games and serious games. *Computers & Education, 59*(2), 661–686.

Denner, J., Werner, L., & Ortiz, E. (2012). Computer games created by middle school girls: Can they be used to measure understanding of computer science concepts? *Computers & Education, 58(*1), 240–249.

Egenfeldt-Nielsen, S. (2006). Overview of research on the educational use of video games. *Digital Kompetanse, 1*(3), 184–213.

Fokides, E. (2018). Digital educational games and Mathematics. Results of a case study in primary school settings. *Education and Information Technologies, 23*(2), 851–867.

Fowler, A. (2012). Enriching student learning programming through using Kodu. *Proceedings of the 3rd Annual Conference of Computing and Information Technology, Education and Research in New Zealand* (incorporating 24th Annual NACCQ).

Garris, R., Ahlers, R., & Driskell, J. E. (2002). Games, motivation, and learning: A research and practice model. *Simulation & Gaming, 33*, 441–467.

Guillén-Nieto, V., & Aleson-Carbonell, M. (2012). Serious games and learning effectiveness: The case of it's a deal! *Computers & Education, 58*(1), 435–448.

Hsiao, H. S., Chang, C. S., Lin, C. Y., & Hu, P. M. (2014). Development of children's creativity and manual skills within digital game-based learning environment. *Journal of Computer Assisted Learning, 30*(4), 377–395.

Hummel, H., Houcke, J., Nadolski, R., Hiele, T., Kurvers, H., & Löhr, A. (2010). Scripted collaboration in serious gaming for complex learning: Effects of multiple perspectives when acquiring water management skills. *British Journal of Educational Technology, 42*(6), 1029–1041.

Kafai, Y. B. (2012). *Minds in play: Computer game design as a context for children's learning.* New York: Routledge.

Kazimoglu, C., Kiernan, M., Bacon, L., & Mackinnon, L. (2012). A serious game for developing computational thinking and learning introductory computer programming. *Procedia-Social and Behavioral Sciences, 47*, 1991–1999.

Ke, F. (2008). A case study of computer gaming for math: Engaged learning from gameplay? *Computers & Education, 51*(4), 1609–1620.

Ke, F. (2014). An implementation of design-based learning through creating educational computer games: A case study on mathematics learning during design and computing. *Computers & Education, 73*, 26–39.

Kirriemuir, J. (2002). *The relevance of video games and gaming consoles to the higher and further education learning experience.* Techwatch Report. Retrieved from https://www.researchgate.net/publication/316856694_The_relevance_of_video_games_and_gaming_consoles_to_the_Higher_and_Further_Education_learning_experience

Kluge, S., & Riley, L. (2008). Teaching in virtual worlds: Opportunities and challenges. *Issues in Informing Science and Information Technology, 5*, 127–135.

Mawer, K., & Stanley, G. (2011). *Digital play: Computer games and language aims*. Peaslake: DELTA Publishing.

McClarty, K. L., Orr, A., Frey, P. M., Dolan, R. P., Vassileva, V., & McVay, A. (2012). *A literature review of gaming in education*. Research Report.

Meerbaum-Salant, O., Armoni, M., & Ben-Ari, M. (2013). Learning computer science concepts with Scratch. *Computer Science Education, 23*(3), 239–264.

Morris, D., Uppal, G., & Wells, D. (2017). *Teaching computational thinking and coding in primary schools*. London: Learning Matters.

Nie, H., Xiao, H. M., & Shang, J. J. (2014, August). A critical analysis of the studies on fostering creativity through game-based learning. *Proceedings of the International Conference on Hybrid Learning and Continuing Education*, 278–287.

Nunes, T., Bryant, P., & Watson, A. (2009). *Key understandings in mathematics learning*. London: Nuffield Foundation.

OfCom, U. K. (2013). *Children and parents: Media use and attitudes report*.

Owston, R., Wideman, H., Ronda, N. S., & Brown, C. (2009). Computer game development as a literacy activity. *Computers & Education, 53*(3), 977–989.

Perrotta, C., Featherstone, G., Aston, H., & Houghton, E. (2013). *Game-based learning: Latest evidence and future directions*. NFER Research Programme: Innovation in Education. Slough: NFER.

Pho, A., & Dinscore, A. (2015). *Game-based learning. Tips and trends*. Retrieved from http://acrl.ala.org/IS/wp-content/uploads/2014/05/spring2015.pdf

Prensky, M. (2001). *Digital game-based learning*. New York: McGraw-Hill.

Robertson, D., & Miller, D. (2009). Learning gains from using games consoles in primary classrooms: A randomized controlled study. *Procedia-Social & Behavioral Sciences, 1*(1), 1641–1644.

Sauvé, L., Renaud, L., & Kaufman, D. (2010). Games, simulations, and simulation games for learning. Definitions and distinctions. In D. Kaufman & L. Sauvé (Eds.), *Educational gameplay and simulation environments: Case studies and lessons learned* (pp. 1–26). Hershey, PA: IGI Global.

Scacchi, W. (2012). *The future of research in computer games and virtual world environments*. Irvine, CA: University of California.

Shokouhi, S., Asefi, F., Sheikhi, B., & Tee, E. R. (2013). Children programming analysis; kodu and story-telling. *Proceedings of the 3rd International Conference on Advance Information System, E-Education & Development*.

Shute, V. J., Rieber, L., & Van Eck, R. (2011). Games . . . and . . . learning. In R. Reiser & J. Dempsey (Eds.), *Trends and issues in instructional design and technology* (3rd ed., pp. 321–332). Upper Saddle River, NJ: Pearson Education Inc.

Spires, H. A. (2008). 21st century skills and serious games: Preparing the N generation. In L. A. Annetta (Ed.), *Serious educational games* (pp. 13–23). Rotterdam: Sense Publishing.

Sung, H. Y., & Hwang, G. J. (2013). A collaborative game-based learning approach to improving students' learning performance in science courses. *Computers & Education, 63*, 43–51.

Susi, T., Johanesson, M., & Backlund, P. (2007). *Serious games-an overview* (technical report). Skövde, Sweden: University of Skövde.

Tobias, S., Fletcher, J. D., Dai, D. Y., & Wind, A. P. (2011). Review of research on computer games. In S. Tobias & J. D. Fletcher (Eds.), *Computer games and instruction* (pp. 127–222). Charlotte, NC: Information Age.

Tüzün, H., Yilmaz-Soylu, M., Karakuş, T., İnal, Y., & Kizilkaya, G. (2009). The effects of computer games on primary school students' achievement and motivation in geography learning. *Computers & Education, 52*(1), 68–77.

Ulicsak, M., & Williamson, B. (2011). *Computer games and learning: A handbook*. London: Futurelab.

Westera, W., Nadolski, R. J., Hummel, H. G. K., & Wopereis, I. (2008). Serious games for higher education: A framework for reducing design complexity. *Journal of Computer Assisted Learning, 24*(50), 420–432.

Chapter 5

Digital learning materials

Could transmedia content make the difference in the digital world?

Polyxeni Kaimara, Sofia-Maria Poulimenou, and Ioannis Deliyannis

Introduction

Given the constant and rapid evolution and thanks to the continuous development of digital technology, the speed of data exchange has increased greatly. Information is immediately transmitted from point to point, while at the same time new daily life skills are deemed necessary to cope with the changes that occur in our everyday lives. New media and multimedia content is already heavily exchanged, a reality that presents learners with new possibilities, one being the introduction of the new digital literacy. Digital literacy requires knowledge construction and qualified learners to share their knowledge (Pietrass, 2007). At the same time, we are all simultaneously learners and trainers, regardless of our age. Social mobility, ubiquitous computing, Internet connectivity, and lifelong learning expand the vision of pedagogy (McLoughlin & Lee, 2008).

Learning concepts including motivation and self-regulation (Schunk, 2012; Zimmerman, 2008), information processing theory (Simon, 1978), socio-cultural learning theory (Vygotsky, 1978), discovery learning theory (Bruner, 1960/1999), and multimedia learning theory (Mayer, 2005) provide the principles on which the pedagogy of the 21st century is based. The broad, easy-to-use, affordable, and daily use of social media interactive computer-mediated technologies have led to the realization that learning is not limited to a single space or a source. There are many perspectives, resources, and learning environments, both real and virtual. Technology has provided us with a communication toolkit that allows anyone to become a journalist and, in theory, on a global scale (McLoughlin & Lee, 2008). Thus, technology enables active experiences and learning outcomes from the world around and promotes the three pedagogical principles: Personalization, participation, and productivity, known as the three P's (McLoughlin & Lee, 2008). Everyone is an active participant and a co-producer rather than a passive consumer of content; he/she is a "prosumer", that is, concurrently producer and consumer. To be a prosumer an active contribution to the content and therefore knowledge construction is essential.

For several decades now, learning theorists have argued that knowledge is not something that is reproduced but something that is being constructed. The revolution brought about by the advent of digital technology in most areas of life is the different quality of interactivity with which one could conduct deeper learning. Deeper and meaningful learning can occur when at least one of three forms of interaction is present at a high level: Student–student, student–instructor, and student–content (Anderson, 2003). An online course designer can substitute one type of interaction for one of the others without degrading the satisfying educational experience and effectiveness. According to Sims (2000), interactivity within the computer-enhanced learning environment is based on four dimensions: Learners, (the who of the learning process), content (the what of the learning process), pedagogy (the how of the learning process), and context (the when and where of the learning process). Learning should be viewed as a complex interaction between circumstances, conditions, environment, motivation, and culture. Changes in knowledge and deeper learning require the acquisition of important skills and competencies by learners, which is a process that begins from early childhood. Deeper learning typically involves sharing what has been learned and interacting with others in a community (Luna Scott, 2015). Therefore, for a deeper understanding of the content, basic pillars are capabilities such as the skills to communicate, collaborate, explore, and create, which are often referred to as "the 4Cs" (NEA, 2012): (a) Critical thinking, (b) communication, (c) collaboration, and (d) creativity. The cultivation of these skills is facilitated by interactive multimedia systems, which are available, affordable, and easily accessible. Interactive multimedia systems highlight the human's leading role by putting him at the centre through cooperative processes that are influenced by the community, culture, and social knowledge (Deliyannis, 2012; Oliver, 2001). Humans interact with the tools designed and developed by others humans, create new ones, decide to be active constructors of knowledge, and share what they create and not the computer does. The optimum framework to describe the interaction between humans and interactive multimedia systems is the Activity Theory (Deliyannis, 2012; Law & Sun, 2012). The Activity Theory is based on Vygotsky's, Luria's and Leont'ev's work, the "troika" of pioneers in what is variously known as the sociocultural, social historical, or cultural-historical theory of development (National Academies of Sciences, Engineering, and Medicine, 2018). In this context, Pedagogy 2.0 (McLoughlin & Lee, 2008) highlights the content that is co-configured by learners themselves through collaborative processes of knowledge construction and cultivation of their critical thinking. Pedagogy 2.0 is ideal for supporting deeper learning because it includes personalized learning strategies, collaborative, and informal learning (Luna Scott, 2015). Such informal learning could be the content of free-choice lifelong learning of cultural heritage activities. The purpose of this chapter is to underline that the content as digital learning material could be the trigger of transforming a student or an ordinary traveller to a cultural heritage content creator. To highlight this content, we approach interactive multimedia and modern devices that support it, based on new pedagogy, which incorporates the fundamental principles of learning theories into ubiquitous learning.

Learning theories in the backdrop of the 21st century pedagogical principles

Back to basics

The 20th-century learning theories were shaped in three theoretical frameworks: Behaviourist learning theory, cognitivist learning theory, and constructivist learning theory (Harasim, 2012). For behaviorism, learning is a process of creating associations between stimuli and responses, aiming at changing the behavior. Skinner, who was one of the most prominent representatives of behaviourism, considered that it is enough to find the right external stimuli and the proper use of rewards alternatively with penalties to achieve the desired behaviour (Kaimara, & Deliyannis, 2019; Skinner, 1938). Learning is enhanced by repetition, especially through the drill and practice process. Behaviourist learning theory has significantly influenced education, guiding the development of highly sequenced and structured curricula, programmed instructional approaches, workbooks, and other tools (Hammond et al., 2001). Behaviourism weaknesses to interpret the learning process as a whole began to be recognized from the early 1920s as insufficient explanations were given for most social behaviours (Harasim, 2012).

Piaget (1929/1971, p. 28) argued that "the child's development is largely the story of the progressive socialization of his individual thought, initially resisting adaptation to social conditions, and then increasingly permeated by adult influences" and he was the first to state that learning is a developmental cognitive process that depends on four factors: Biological maturation, experience with the physical environment, experience with the social environment, and equilibration, that is, the balance between assimilation and accommodation (Schunk, 2012). Several variations in the cognitive approach are developed, such as the information processing theory (Simon, 1978). In the 1960s, the information processing theory began to replace the behaviourism that prevailed until then in psychology. Mayer (1996) argued that the information processing theory and constructivism may be considered as approaches to cognitive psychology. In cognitivist and constructivist approaches the learning phenomenon is a process that is constructed by the action of the learner himself rather than as a result of stimulus–response associations, which is the fundamental axiom of behaviourism.

The active role of learners was also emphasized by Vygotsky (1978) who was deeply interested in the role of the social environment, included tools and cultural objects and thus introduced the social aspect of learning into constructivism. The concept of mediation in the human–environment interaction to the use of signs as well as tools and systems (language, writing, number systems), artefacts such as technology and its use is determined socially, culturally, and historically (Sannino et al., 2009; Uden & Helo, 2008). "Tools are created by societies over the course of human history and change with the form of society and the level of its cultural development" (Vygotsky, 1978, p. 7). The social relevance and the impact of activity theory depend on our ability to understand the changing nature of the objects. The evolution of the activity theory can be conceived in terms of three generations: The first generation is based on the concept of Vygotsky's mediation, the second generation is based on the concept of Leont'ev's activity system and the third generation is based on the idea of multiple interacting activity systems focused on a partially shared object (Engeström, 2009). Bruner (1961) who was increasingly influenced by Lev Vygotsky, stated that if one is able to approach learning as a task of "discovering something" rather than "learning about it" then he/she carries out his/her activity with the autonomy of self-reward. His theory, known as the discovery learning theory, highlighted the child's position to experience success or failure, not like reward or punishment as behaviourists had suggested, but instead as information. Even if theorists of cognitivist and constructivist approaches differed in important ways, they all shared an emphasis on considering children as active learners (Bransford et al., 2000). Humans are viewed as goal-directed agents who actively seek information with previous knowledge, skills, beliefs, and concepts (Kaimara & Deliyannis, 2019).

What do all these conclusions furnish in an era characterized by the rapid development of technology? According to Dede (2008), the historical controversy between pedagogy and technology outlines the necessity of a universal educational method that is best for all types of content, students, and educational goals based on the composition of the different principles of learning theories. Behaviourism, cognitivism, and constructivism are the three learning theories most often utilized in creating learning environments. These theories, however, were developed at a time when learning was not influenced by digital technology. Behaviourism and cognitivism consider knowledge as external to the learner and the learning process is an action that internalizes knowledge. On the other hand, constructivism doesn't regard the learner as an empty vessel to be filled with knowledge (Siemens, 2005). Nowadays, technology reorganizes our lives, our communication, and ultimately our learning. Learning needs require theories supported by principles that reflect the contemporary social environment.

Designing learning for the 21st century

Traditional learning theories consider learning as a process of acquiring and reorganizing cognitive structures within the closed boundaries of specific tasks. In traditional teaching methods, teachers and students follow a curriculum (syllabus) aimed at a fantastic medium representative learner. The traditional type of learning is incapable of responding to the demands of complex social change and the creation of novel artefacts and social structures (Engeström, 2009). Educators have adopted new technologies, but still, largely through trial-and-error methods and by adapting traditional didactic practices to online environments, both within formal and nonformal educational settings (Harasim, 2012). Nevertheless, successful educational systems worldwide depend on transforming pedagogy and redesigning learning tasks (Leadbeater, 2008). Lee's and Tsai's (2011) research showed significant differences in students' perceptions between Internet-based learning and traditional learning in terms of collaboration, self-regulated learning, and information seeking. Self-regulation of learning involves more than the detailed knowledge of a skill. Self-regulation premises the self-awareness, self-motivation, and behavioral skill to implement that knowledge appropriately (Zimmerman, 2008).

For over a century, Dewey has emphasized that in traditional educational systems, greater importance is given to the material to be studied and not to the learner's experience and thus "the act of learning is artificial and ineffective in the degree in which pupils are merely presented with a lesson to be learned" (Dewey, 1916/1922, p. 158). While outside school, students interact with natural facts and principles related to various manifestations of human action and therefore strive to understand the material and processes involved, inside the school environment they are forced to break the continuity of mental development. The learning material is isolated from his/her daily experiences and students feel that the social world is unrelated to their courses leading to a lack of motives and interests. For Dewey (1938/1986), although young people in traditional schools obtain experiences, the problem is related to their quality and their defective and wrong character: Wrong and defective from the standpoint of connection with further experience. Bybee and his colleagues inspired from several instructional models, approaches, philosophers, and science teachers, including Herbart, Dewey, Hawkins, Piaget, Atkin, and Karplus (Bybee et al., 2006; Bybee, 2009) proposed the Biological Sciences Curriculum Study (BSCS) 5E instructional model. Although BSCS 5E was developed as an instructional model for improving science education, it is now adapted and used to improve instruction in another area, including technology education and mathematics. This model provides a method of helping learners move progressively towards stronger understanding given the opportunity to discover meaning by themselves through engagement, exploration, explanation, elaboration, and evaluation. The 5E model provides a tangible reference for teachers to scaffold their developing expertise in structuring a learning environment that will facilitate students' interaction (Boddy, Watson & Aubusson as cited in Senan, 2013). Senan (2013) suggested that the 5E model can aid teachers in structuring learning experiences by designing a multimedia learning package.

Simply put, multimedia learning (Mayer, 2005) constructs mental representations from words (such as the spoken or printed text) and pictures (such as illustrations, photos, animation, or video). According to the literature review (Mayer, 2005, 2014), cognitive multimedia learning theory proposes 12 basic (Table 5.1) and 9 advanced (Table 5.2) principles for designing multimedia learning environments.

Multimedia learning principles serve to optimize the cognitive load on the learner as humans are constantly facing challenges and new experiences

Table 5.1 Basic principles of multimedia learning

1	Multimedia principle	People learn better from words and pictures than from words alone
2	Modality principle	People learn better from graphics and narrations than from animation and on-screen (printed) text (people learn better from a multimedia message when the words are spoken rather than written)
3	Redundancy principle	People learn better when the same information is not presented in more than one format (however, there are special situations in which the related audio might be difficult for the learner to understand, as with foreign language learning or certain auditory learning disabilities)
4	Segmenting principle	People learn better when a multimedia lesson is presented in user-paced segments rather than as a continuous unit
5	Pre-training principle	People learn better from a multimedia lesson when they know the names and characteristics of the main concepts
6	Coherence principle	People learn better when extraneous material is excluded rather than included
7	Signaling principle	People learn better when cues that highlight the organization of the essential material are added
8	Spatial contiguity principle	People learn better when corresponding words and pictures are presented near rather than far from each other on the page or screen
9	Temporal contiguity principle	People learn better when corresponding words and pictures are presented simultaneously rather than successively
10	Personalization principle	People learn better when the words of a multimedia presentation are in conversational style rather than formal style
11	Voice principle	People learn better when the words are spoken in a standard-accented (friendly) human voice rather than a machine voice or foreign-accented human voice
12	Image principle	People do not necessarily learn better when the speaker's image is on the screen

Table 5.2 Advanced principles of multimedia learning

1	Guided-discovery principle	People learn better when guidance is incorporated into discovery-based multimedia environments
2	Worked-out example principle	People learn better when they receive worked-out examples in initial skill learning
3	Collaboration principle	People can learn better with collaborative online learning activities
4	Self-explanation principle	People learn better when they are encouraged to generate self-explanations during learning
5	Animation and interactivity principles	People do not necessarily learn better from animation than from static diagrams
6	Navigation principles	People learn better in hypertext environments when appropriate navigation aids are provided
7	Site map principle	People can learn better in an online environment when the interface includes a map showing where the learner is in the lesson
8	Prior knowledge principle	Instructional design principles that enhance multimedia learning for novices may hinder multimedia learning for more expert learners
9	Cognitive ageing principle	Instructional design principles that effectively expand working memory capacity are especially helpful for older learners

(Berger & Krousgrill, 2012). Indeed, Siemens (2005) is among all of them who argued that experience has long been seen as the best teacher of knowledge. Since we cannot experience everything, the experiences of other people become a substitute for knowledge. Nowadays, it is possible for students to communicate with other students who may be in another country and thus to have feedback, reflection, and review. Siemens coined the term "connectivism" to describe a new learning theory. Although there is a debate about whether the connectivism is a real learning theory or an instructional theory or simply a pedagogical view (Barry, 2013; Clarà & Barberà, 2014; Duke et al., 2013; Kop & Hill, 2008), Siemens (2005, 2008) himself argued that the previous theories, even social constructive views, assume that learning occurs within the individual and connectivism is the new framework in which the current trends of learning are described. It is recognized that connectivism provides four pedagogical characteristics of massive online open courses (MOOC), that is, diversity, autonomy, interactivity or connectedness, and openness (Downes, 2010) and eight pedagogical principles fully assumable by a large number of learning theories (Siemens, 2005):

- Learning and knowledge rest in a diversity of opinions
- Learning is a process of connecting specialized nodes or information sources

- Learning may reside in non-human appliances
- Learning is more critical than knowing
- Nurturing and maintaining connections is needed to facilitate continual learning
- Ability to see connections between fields, ideas, and concepts is a core skill
- Currency (up-to-date knowledge) is the intent of all connectivist learning activities
- Decision-making is itself a learning process

All these new trends premise a person who is an active explorer of his/her environment, a person who has self-awareness, who acts, experiments, experiences, belongs, and interconnects. Siemens (2008) noted that the use of participative technologies for communication and collaboration is increasingly common.

In such a networked society, McLoughlin and Lee (2008) suggested that learning should be based on three pedagogical principles: Personalization, participation, and productivity (e.g. content creation), known as the 3 P's. Personalized learning inspires teachers to design ways of approaching students individually, adjusting to a particular learner's interests, goals, focus, progress, and problems (Spector, 2014) following the principles of differentiated instruction that supports learners with learning difficulties or special educational needs, learners with different mother language, cultural, religious, or/and socio-economic background, putting the bases for inclusive education (Kaimara et al., 2019, Tomlinson et al., 2003). Spector (2016) underlined the difference between "personal" and "personalized" and stressed that the "personal learning environment" constructed and controlled by the learner is completely different from the "personalized learning environment" that automatically adapts to the interests and needs of individual learners. The contribution of social media to learners' desires to participate and connect with others is catalytic and, at the same time, the dynamics of these media to support personal meaningful learning through collaboration and shared knowledge building are reflected (McLoughlin & Lee, 2008). Today's young people increasingly see themselves as participants in creating information and ideas, as members of social networks, not merely spectators. They also perform the roles of editing, revising, and organizing the content, which becomes part of the shared pool of resources accessible to all learners, i.e. wiki. While they are playing games online, peer-to-peer collaboration is encouraged (Leadbeater, 2008; McLoughlin & Lee, 2008), and while they are creating and participating in the Web discussions, evaluating content of and for other users, sharing applications, and publishing their content become "prosumers", and the third of the pedagogical principles of 21st century "productivity" is fulfilled. But what skills should a person need to be able to develop to gain personalized learning and at the same time participate and be productive?

The 4Cs super skills

Nowadays, students move beyond the basics, i.e. *r*eading, w*r*iting, a*r*ithmetic (3 Rs model) and embrace the 4Cs, the "super skills" for the 21st century: Critical thinking, communication, collaboration, and creativity (Kivunja, 2015; NEA, 2012). These skills should be considered as interdependent and interrelated and only as indicative of the skills that a citizen needs of a networked society. Certainly, critical thinking, communication, collaboration, and creativity are not new or innovative for experts, educators, and policy-makers on the development of educational issues and systems. Humans need to think critically, communicate, collaborate, and create. Nevertheless, in a society that no one can predict its technological evolution, each current cutting-edge device soon becomes old with fewer features than the next one and each citizen is called upon to adapt and learn and, above all, to benefit from the new technological capabilities. We live in a society that is constantly learning and when 4Cs are combined with technological progress, learning methods are expanding and undoubtedly an evolutionary step towards the development of super skills is achieved.

Critical thinking

Critical thinking is integral to education and rationality and is traceable to the teaching practices of Socrates of ancient Greece (National Council for Excellence in Critical Thinking, n.d.). If teachers want their students to think well, they must help students understand the essentials of thought and reasoning: Purpose, question, problem-solving, assumptions, point of view, data, information, evidence, concepts, ideas, inferences or interpretations, conclusions, implications, and consequences. These elements construct thinking, thus the individual's ability to use general cognitive processing skills. Critical thinking combines with Bloom's cognitive taxonomy and its revision (Krathwohl, 2002), which focuses on an individual's ability to remember, understand, apply, analyse, evaluate, and construct new ideas. Currently, people are continuously facing new challenges and new experiences. In an era of rapid communication and transfer of information, humans interact with people and computers and have to manipulate a lot of data. As a consequence, there is so much information for working memory to handle; therefore, there is a risk of cognitive overload. On one hand, digital literacy is required, and on the other hand, skills and competences are in demand to ensure the offered and available information is real and useful, for it to then become knowledge.

Communication

Communication is about understanding and sharing thoughts, questions, ideas, and solutions. In a global society that is characterized by the rapid spread of information, media, and digital technological development, the necessity for effective communication becomes more apparent (Kivunja, 2015). In a school context, effective teaching and learning processes depend on successful communication, during teachers' and students' interaction. The cultivation of interactive and transactional communication skills is essential for students' success, not only in the academics domain but also in lifelong activities. These skills are formulated in verbal, non-verbal, written, audio, visual, or digital communication. There has been increasing interest in examining the role of social media including asynchronous and synchronous online communication such as email, chat rooms, instant messages, text messages, web-based social networking sites, cooperative multiplayer games, and video-sharing sites in adolescent development. Some studies have found that Internet use can be a solitary and socially isolating activity, while others have argued that social media use can enhance social relationships among both adults and adolescents. The adolescents interacted in online communication largely with existing circles of friends and reported feeling closer to them when they communicated online (Mazurek & Wenstrup, 2013). A remarkable case of communication is that of people with the autism spectrum disorder (ASD). People with ASD appear to have deficits in the social-affective and behavioural domains and often difficulties in joint attention, non-verbal communication, the production of referential language, and various other social and communications skills. There is evidence that assistive technology and video modelling is more effective than in-vivo modeling in teaching social skills. A possible explanation could be that video modeling procedures are more motivating and raise attention because stimulus novelty increases attention and enhances observational learning (Charlop et al., 2018). Additionally, asynchronous online communication offers opportunities for social interaction among individuals with ASD, in a format that does not require attention to non-verbal cues, gestures, facial expressions, or vocalization (Mazurek & Wenstrup, 2013). Social networking technology offers users the ability to adjust their social interaction timeframe by exchanging digitized multimedia content and experiences within their social circle at their own pace and location (Deliyannis, 2012).

Collaboration

Vygotsky (1978) argued that learning and cognitive development are culturally and socially based. Learning is not happening in isolation, it is not a process that matters only to an individual, but it happens through the dynamic interactions among individuals. In other words, learning is a social process and occurs during interactions and collaborations between individuals. Nowadays, in a digitally networked society, online collaborations come about providing opportunities for both teachers and students to interact. Collaborative learning is the process in which students develop higher-order thinking skills by creating an environment where knowledge is shared among learners in an effort to achieve common learning goals (Brindley et al., 2009). Dewey (1916/1922, p. 188) could say that "in such shared activity, the teacher is a learner, and the learner is, without knowing it, a teacher- and upon the whole, the less consciousness there is, on either side, of either giving or receiving instruction, the better". Examples of online collaboration could be the Web-based Google doc, wiki, videoconferences, portable interactive whiteboards with free downloads applications that record pen strokes, and audio simultaneously.

Creativity

Vygotsky (2004) defined creativity as one of the most important areas of a child's general development and maturation and a significant issue of educational psychology. It showed the relation of creativity to imagination through characteristic examples of authentic and truest creativity, during the spontaneous play of children: When a child is sitting astride a stick and is pretending to be riding a horse; as a girl is playing with her doll, she is imagining she is its mother; a boy who becomes a pirate, a soldier or a sailor, all these children at play represent examples. He emphasized the particular importance of cultivating creativity in school-age children. However, he underlined that

> when we attempt to foster children's creativity, including in the visual arts, we need to observe the principle of freedom, which is generally an essential condition for all kinds of creativity. This means that the creative activities of children cannot be compulsory or forced and must arise only out of their own interests.
>
> (Vygotsky, 2004, p. 84)

The child's interest differs from an adult's interest and thus a child's imagination functions differently from an adult's. Vygotsky's views about creativity are back on track as it has become more of a necessity in today's information, technology, and innovation-driven society and that is because there is a plethora of creative and communication aspects offered by interactive multimedia (Deliyannis, 2012). Active users in social networks perceive interactive multimedia applications as a tool that allows them to be informed virtually in contrast to passive media such as TV. Creativity transforms both the creator through personal experience and other people via the creation of knowledge and innovative artifacts. Creativity is both the goal and the means of personal and cultural development (Moran et al., 2003).

Acquiring knowledge of cultural heritage in a networked society

Learning is a continuous process and it is not only about students of childhood and adolescents but a lifelong process. Anyone can be a potential learner because learning can be formal, non-formal, and informal (Colardyn & Bjornavold, 2004). Formal learning is integrated into the educational system, for both minors and adult learners, and can take place inside or outside the classroom, while non-formal learning is related to planned activities that are not strictly defined as learning but contain learning elements. The concept of informal learning is a process resulting from daily life activities taking place at any time, in any place, and in any context, e.g. work, family, or leisure. The researchers concluded that the time the children spent in structured learning activities limited their opportunities to learn, to experience, to control themselves in natural and informal learning contexts. These informal contexts are critical for effective learning in the real world (National Academies of Sciences, Engineering, and Medicine, 2018). There are courses of formal education such as history and geography that, when taught in the traditional way, lead students to boredom. In teaching material, there is a plethora of dates, names of cities, and people and events, which children are asked to memorize. To create links between places, events, and people, teachers are equipped with maps or pictures or documentaries, at best. Probably, they will connect the course with an excursion to a museum, which most of the time, especially in primary education, is just a chance for a tour. From this perspective, museums and sites of interest began to transform. Keeping up with new progress, these sites are being modernized and start to offer guided tours in a more interactive way. Thus, a course in a museum doesn't look like a real course. The role of technology in informal learning—and the potential for linking it to formal learning—have only become more salient, as daily life is increasingly mediated by digital and Internet technology. Museums and arts programmes are ideal where technology-supported activities combine learning with entertainment (National Academies of Sciences, Engineering, and Medicine, 2018). The designs of most contemporary museums are based on the dynamic principles of constructive theories. The audience, adults and minors, can enter data; visitors can answer questions or even solve puzzles; or play in rival teams. Using their own smartphones, visitors can target an exhibit and can get more information through augmented reality apps. Guests are also given the opportunity to add their own information. This interaction acquires elements of game experience. Thus, everyday gadgets such as the smartphone-tablets are transforming into tools for digital learning materials. Smart devices have the capability to support and engage their owners anytime and anywhere and thus can improve learning, sometimes in a playful way.

In this framework, the Department of Audiovisual Arts of Ionian University, Corfu, Greece, has designed an ongoing project dedicated to the 4Cs "critical thinking, communication, collaboration and creativity", that creates a methodology for accessing different levels of information about a United Nations Educational, Scientific and Cultural Organization (UNESCO) World Heritage Site. The project is being implemented at the Old Town of Corfu, a recognized historical monument by the Greek Law since 1980

(Presidential decree, 1980) and a UNESCO World Heritage site (decision 31 COM 8b.40) since 2007 (UNESCO, 2007). The project idea deploys 360° video technology, augmented reality, and holograms to offer unlimited access to information to the visitors and users of a monument with the application of an interactive map with pre-selected points of interest placed on it, where it is possible to create and conduct interactive tours for different user categories. The interactive tours provide cultural and historical content for every single point according to each user's needs and interest. At the same time, relevant stakeholders such as researchers, professors, students, local authorities, tourism, and cultural enterprises, are able to enrich the system with new content linked to the points of interest, creating new thematic routes, and personalized narration.

The implementation methodology consists of four main activities that enhance the learning experience through:

A The creation of an easy-to-use application for mobile communication devices that provides high-quality information and multimedia content for the points of interest.
B The creation of holographic content in the form of a video about the Old Town of Corfu, where users may interact with 3D points of interest by focusing, rotating, and exploring in detail a holographic display.
C The creation of a Geospatial Information System for the above information for remote users in its full audiovisual form, including 360° video information for tourism and educational purposes.
D The creation of a Thematic Channel about the Old Town of Corfu connected to the EUROPEANA Platform with full free access to uploaded metadata for their re-use and exploitation.

The above-mentioned hologramatic display includes the 3d recreation of very important historic buildings in the Old Town of Corfu, with the goal of achieving optimum photorealism. For this cause, the reconstruction of the buildings and their surrounding environment uses sketches, photos from the past, current high-resolution pictures of the real building, and drone videos to reach a level of detail as close as possible to the original.

The following image (Figure 5.1) displays the initial reconstruction of the Town Hall of Corfu, which is the former San Giacomo Theatre, a 16th-century building and one of the most historic places in Corfu.

The main goal of the project is to support the documentation and protection of cultural heritage and at the same time to promote the learning process for both communication and educational purposes and to enhance the creation of effective outcomes, like empathy, which are the fundamental aspects of the learning experience (Bachen et al., 2016).

On the other hand, the main innovation of the project lies in the combined use of different, modern, multi-sensory technologies in a way that adapts

Figure 5.1 **Image by Leonidas Koufalis, 3D designer and Nikolaos Mamalos, photographer, external experts at the project "Hologramatic Corfu", with the task of recreating parts of the Old Town of Corfu.**

to the needs of users. In particular, the project emphasizes on 3D digital animated holograms, where the users can perceive and actively engage with the cultural content and be effectively moved by the experience as they can enjoy the monument as a whole, looking at details that would be in no other way accessible (e.g. the roof of a building). These accurate and immersive 3D recreations of a monument function as facilitating factors for learning in non-formal and informal contexts (Mortara & Catalano, 2018).

The users of the system act at the same time as learners, while exploring the information given and as knowledge creators and actors, since they enhance the system with their own information, generating new content for others to attain. In this way, each user of the system may create their own personalized content about a place and enjoy self-related customized knowledge.

Furthermore, the users may communicate with each other with an online community created by registered users of the system that allows anyone to become a Site Ambassador, offer their knowledge about a place and promote the learning experience in a creative, communicative, and collaborative way. Also, the ability of the system to transform a user's experience into new content enhances critical thinking as the user's skill for further knowledge acquirement.

At the moment the project has only been piloted in a small number of cultural settings in the Old Town of Corfu and is currently in progress as an overall application.

Conclusion

The role of pedagogy becomes more important for finding ways to integrate technology in education (Daniela, 2019). Educational technology can engage learners and educators and can facilitate the implementation of the new educational philosophy of the 21st century. This new educational philosophy focuses not on the reproduction of knowledge but its construction. The 21st century pedagogical principles incorporate elements such as personalization, participation, and productivity. Educators, in formal and informal environments, to help their learners to cope with the new challenges find ways to cultivate critical thinking, communication, collaboration, and creativity. In this context, a new term for pedagogy emerged: Smart pedagogy. Smart pedagogy has the capability to adapt the fundamentals of previous learning theories in such a way to fit with the new challenges. Our environment, natural and man-made, steadily provides stimuli. The majority of people are constantly carrying modern smart devices. We often see young people sitting side by side and not communicate face to face but through smart devices. The unsuspecting observer imagines that communication between young people has disappeared while the more aware thinks that communication has simply changed form and means. In smart pedagogy, the "smart" teacher can exploit young people's behaviour and use their devices as triggers for creation, participation, personalization, critical thinking, communication, and collaboration and ultimately for learning. In fact, the educator can turn the devices into educational tools and motivate not only the young but also elderly people to interact with the content. The introduction of technological tools, the transformation of any environment into learning and the exploitation of every occasion, anytime and anywhere, leads to knowledge building and the interaction with social networks changes society into a continuously learning one (Daniela, 2019). Transmedia learning and handheld devices reflect the new pedagogy towards ubiquitous learning and allow the continuous supply of content and information from distant sources by the active participants in a learning society. Traditionally, to teach cultural heritage-oriented content is a boring procedure. The dynamics of new interactive media support the cultural heritage-oriented content and engage learners giving them the control of the process, through interaction with the content and other content creators.

References

Anderson, T. (2003). Getting the mix right again: An updated and theoretical rationale for interaction. *International Review of Research in Open and Distance Learning, 4*(2). doi:10.19173/irrodl.v4i2.149

Bachen, C. M., Hernández-Ramos, P. F., Raphael, C., & Waldron, A. (2016). How do presence, flow, and character identification affect players' empathy and interest in learning from a serious computer game? *Computers in Human Behavior, 64*, 77–87.

Barry, W. (2013). Connectivism: Theory or phenomenon? *The Accidental Technologist.* Retrieved from http://www.waynebarry.com/2013/04/29/connectivism-theory-or-phenomenon

Berger, E. J., & Krousgrill, C. M. (2012). HigherEd 2.0: Web 2.0 in higher education. In I. Deliyannis (Ed.), *Interactive multimedia* (pp. 67–92). Rijeka: IntechOpen. doi: 10.5772/37505

Bransford, J. D., Brown, A., & Cocking, R. (2000). *How people learn: Mind, brain, experience, and school.* Washington, DC: National Research Council. doi:10.17226/9853

Brindley, J., Blaschke, L. M., & Walti, C. (2009). Creating effective collaborative learning groups in an online environment. *The International Review of Research in Open and Distributed Learning, 10*(3). doi:10.19173/irrodl.v10i3.675

Bruner, J. (1960/1999). *The process of education.* Cambridge, MA:Harvard University Press (Original work published 1960).

Bruner, J. S. (1961). The act of discovery. *Harvard Educational Review, 31*(1), 21–32. Retrieved from https://digitalauthorshipuri.files.wordpress.com/2015/01/the-act-of-discovery-bruner1.pdf

Bybee, R. W. (2009). *The BSCS 5E instructional model and 21st century skills.* Colorado Springs, CO: BSCS.

Bybee, R. W., Taylor, J. A., Gardner, A., Van Scotter, P., Powell, J. C., Westbrook, A., & Landes, N. (2006). *The BSCS 5E instructional model: Origins and effectiveness.* Colorado Springs, CO: BSCS.

Charlop, M. H., Lang, R., & Rispoli, M. (2018). Lights, camera, action! teaching play and social skills to children with autism spectrum disorder through video modeling. In *Play and social skills for children with autism spectrum disorder* (pp. 71–94). Cham: Springer. doi:10.1007/978-3-319-72500-0_5

Clarà, M., & Barberà, E. (2014). Three problems with the connectivist conception of learning. *Journal of Computer Assisted Learning, 30*(3), 197–206. doi:10.1111/jcal.12040

Colardyn, D., & Bjornavold, J. (2004). Validation of formal, non-formal and informal learning: Policy and practices in EU Member States1. *European Journal of Education.* doi:10.1111/j.0141-8211.2004.00167.x

Daniela, L. (2019). Smart pedagogy for technology-enhanced learning. In *Didactics of smart pedagogy* (3–21). Cham: Springer. doi:10.1007/978-3-030-01551-0_1

Dede, C. (2008). Theoretical perspectives influencing the use of information technology in teaching and learning. In J. Voogt & G. Knezek (Eds.), *International handbook of information technology in primary and secondary education.* (pp. 43–62). Boston, MA: Springer. doi:10.1007/978-0-387-73315-9_3

Deliyannis, I. (2012). From interactive to experimental multimedia. In I. Deliyannis (Ed.), *Interactive multimedia.* (pp. 3–12). Rijeka: IntechOpen. doi:10.5772/38341

Deliyannis, I., & Kaimara, P. (2019). Developing smart learning environments using gamification techniques and video game technologies. In L. Daniela (Ed.), *Didactics of smart pedagogy* (pp. 285–307). Cham: Springer.doi:10.1007/978-3-030-01551-0_15

Dewey, J. (1916/1922). *Democracy and education: An introduction to the philosophy of education.* New York: Macmillan.

Dewey, J. (1938/1986). Experience and education. *The educational forum, 50*(3), 241–252. Taylor & Francis Group.

Downes, S. (2010). Learning networks and connective knowledge. In *Collective intelligence and E-Learning 2.0: Implications of web-based communities and networking* (pp. 1–26). IGI Global. doi:10.4018/978-1-60566-729-4.ch001

Duke, B., Harper, G., & Johnston, M. (2013). Connectivism as a digital age learning theory. *The International HETL Review, 2013*(Special Issue), 4–13.

Engeström, Y. (2009). The future of activity theory: A rough draft. In A. Sannino, H. Daniels, & K. D. Gutiérrez (Eds.), *Learning and expanding with activity theory* (pp. 303–328). Cambridge University Press. dx:10.1017/CBO9780511809989.020

Hammond, L-D., Austin, K., Orcutt, S., & Rosso, J. (2001). *How people learn: Introduction to learning theories*. Stanford University. Retrieved from http://web.stanford.edu/class/ed269/hplintrochapter.pdf.

Harasim, L. (2012). *Learning theory and online technologies*. Routledge. doi: 10.4324/9780203846933

Kaimara, P., & Deliyannis, I. (2019). Why should I play this game? The role of motivation in smart pedagogy. In L. Daniela (Ed.), *Didactics of smart pedagogy* (pp. 113–137). Cham, Springer. doi:10.1007/978-3-030-01551-0_6

Kaimara, P., Poulimenou, S., Oikonomou, A., Deliyannis, I., & Plerou, A. (2019, April). Smartphones at schools? Yes, why not?. *European Journal of Engineering Research and Science*. CIE, 1–6. doi:10.24018/ejers.2019.0.CIE.1288

Kivunja, C. (2015). Exploring the pedagogical meaning and implications of the 4Cs "super skills" for the 21st century through Bruner's 5E lenses of knowledge construction to improve pedagogies of the new learning paradigm. *Creative Education, 6*(02), 224.

Kop, R., & Hill, A. (2008). Connectivism: Learning theory of the future or vestige of the past? *The International Review of Research in Open and Distributed Learning, 9*(3). doi:org/10.19173/irrodl.v9i3.523

Krathwohl, D. R. (2002). A revision of Bloom's taxonomy: An overview. *Theory into practice, 41*(4), 212–218.

Law, E. L. C., & Sun, X. (2012). Evaluating user experience of adaptive digital educational games with activity theory. *International Journal of Human-Computer Studies, 70*(7), 478–497. doi:10.1016/j.ijhcs.2012.01.007

Leadbeater, C. (2008). What's next? 21 ideas for 21st century learning. London: The Innovations Unit. Retrieved from http://charlesleadbeater.net/wp-content/uploads/2010/01/Next2.pdf

Lee, S. W. Y., & Tsai, C. C. (2011). Students' perceptions of collaboration, self-regulated learning, and information seeking in the context of Internet-based learning and traditional learning. *Computers in Human Behavior, 27*(2), 905–914.

Luna Scott, C. (2015). The futures of learning 2: What kind of learning for the 21st century? *UNESCO Education Research and Foresight Working Papers Series*. Retrieved 1 March 2019 from http://unesdoc.unesco.org/images/0024/002429/242996E.pdf

Mayer, R. E. (1996). Learners as information processors: Legacies and limitations of educational psychology's second. *Educational psychologist, 31*(3–4), 151–161.

Mayer, R. E. (2005). Introduction to multimedia learning. In *The Cambridge handbook of multimedia learning* (pp. 1–16). New York: Cambridge University Press. doi:10.1017/CBO9780511816819.002

Mayer, R. E. (2014). *The Cambridge handbook of multimedia learning* (2nd ed.). New York: Cambridge University Press. doi:10.1017/CBO9781139547369.002

Mazurek, M. O., & Wenstrup, C. (2013). Television, video game and social media use among children with ASD and typically developing siblings. *Journal of Autism and Developmental Disorders, 43*(6), 1258–1271. doi:10.1007/s10803-012-1659-9

McLoughlin, C., & Lee, M. J. (2008). The three p's of pedagogy for the networked society: Personalization, participation, and productivity. *International Journal of Teaching and Learning in Higher Education, 20*(1), 10–27.

Moran, S., John-Steiner, V., & Sawyer, R. K. (2003). Creativity in the making. Vygotsky's contemporary contribution to the dialectic of development and creativity. In R. K. Sawyer, V. John-Steiner, S. Moran, R. J. Steinberg, D. H. Feldman, J. Nakamura, & M. Csikszentmihalyi (Eds.). *Creativity and development* (pp. 61–90). New York: Oxford University Press. doi:10.1093/acprof:oso/9780195149005.003.0003

Mortara, M., & Catalano, C. (2018). 3D Virtual environments as effective learning contexts for cultural heritage. *Italian Journal of Educational Technology, 26*(2), 5–21. Ortona: Edizioni Menabò – Menabò srl. Retrieved from https://www.learntechlib.org/p/195252/

National Academies of Sciences, Engineering, and Medicine. (2018). *How people learn II: Learners, contexts, and cultures.* Washington, DC: The National Academies Press. doi:10.17226/24783

National Council for Excellence in Critical Thinking. (n.d.). *Defining critical thinking.* Retrieved from http://www.criticalthinking.org/pages/the-national-council-for-excellence-in-critical-thinking/406

National Education Association (NEA). (2012). Preparing 21st century students for a global society: An educator's guide to the four C's. Retrieved from https://bit.ly/2VB7k6p

Oliver, R. (2001). Developing e-learning environments that support knowledge construction in higher education. In S. Stoney & J. Burn (Eds.), *Working for excellence in the e-conomy.* (pp 407–416). Churchlands: Australia,We-B Centre.

Piaget, J. (1929/1971). *The child's conception of the world.* London: Routledge & Kegan Paul.

Pietrass, M. (2007). Digital literacy research from an international and comparative point of view. *Research in Comparative and International Education, 2*(1), 1–12. doi:10.2304/rcie.2007.2.1.1

Presidential decree (22-04-1980). For the characterization of a specific area of the Old Town of Corfu as a traditional landscape (number 274/D/1980 of the Official Government Gazzette of the Hellenic Parliament, issued on 05-05-1980)

Sannino, A., Daniels, H., & Gutiérrez, K. D. (2009). Activity theory between historical engagement and future-making practice. In A. Sannino, H. Daniels, & K. D. Gutiérrez (Eds.), *Learning and expanding with activity theory* (pp. 1–15). Cambridge University Press. doi:10.1017/CBO9780511809989.002

Schunk, D. H. (2012). *Learning theories: An educational perspective* (6th ed.). Boston, MA: Pearson Education Inc.

Senan, D. C. (2013). Infusing BSCS 5E instructional model with multimedia: A promising approach to develop 21st century skills. *Journal on School Educational Technology, 9*(2), 1–7.

Siemens, G. (2005, January). Connectivism: A learning theory for a digital age. *International Journal of Instructional Technology and Distance Learning, 2*(1). Retrieved from http://www.itdl.org/Journal/Jan_05/article01.htm

Siemens, G. (2008). Learning and knowing in networks: Changing roles for educators and designers. *ITFORUM for Discussion, 27*, 1–26.

Simon, H. A. (1978). Information-processing theory of human problem solving. *Handbook of Learning and Cognitive Processes, 5*, 271–295.

Sims, R. (2000). An interactive conundrum: Constructs of interactivity and learning theory. *Australasian Journal of Educational Technology, 16*(1). doi:10.14742/ajet.1821

Skinner, B. F. (1938). *The behavior of organisms: An experimental analysis.* New York: Appleton Century.

Spector, J. M. (2014). Conceptualizing the emerging field of smart learning environments. *Smart Learning Environments, 1*(1), 2. doi:10.1186/s40561-014-0002-7

Spector, J. M. (2016). The potential of smart technologies for learning and instruction. *International Journal of Smart Technology and Learning, 1*(1), 21–32. doi:10.1504/IJSMARTTL.2016.078163

Tomlinson, C. A., Brighton, C., Hertberg, H., Callahan, C. M., Moon, T. R.B., Brimijoin, K., Conover, L. A., & Reynolds, T. (2003). Differentiating instruction in response to student readiness, interest, and learning profile in academically diverse classrooms: A review of literature. *Journal for the Education of the Gifted, 27*(2–3), 119–145. doi:10.1177/016235320302700203

Uden, L., & Helo, P. (2008). Designing mobile interfaces using activity theory. *International Journal of Mobile Communications, 6*(5), 616–632. doi:10.1504/IJMC.2008.019325

United Nations Educational, Scientific and Cultural Organization – UNESCO (2007). Old Town of Corfu. Retreived from https://whc.unesco.org/en/list/978/

United Nations Educational, Scientific and Cultural Organization – UNESCO, World Heritage site (decision 31 COM 8b.40) *Nomination of natural, mixed and cultural properties to the world heritage list - Old Town of Corfu,* Retreived from https://whc.unesco.org/en/decisions/1339

Vygotsky, L. S. (1978). *Mind in society: The development of higher psychological processes.* (M. Cole, V. John-Steiner, S. Scribner, & E. Souberman, Eds.). Cambridge, MA: Harvard University Press.

Vygotsky, L. S. (2004). Imagination and creativity in childhood. *Journal of Russian & East European Psychology, 42*(1), 7–97.

Zimmerman, B. J. (2008). Investigating self-regulation and motivation: Historical background, methodological developments, and future prospects. *American Educational Research Journal, 45*(1), 166–183.

Chapter 6

Unpacking digital play in children's play

Jennifer Wong-Powell

Introduction: Digital technologies in children's play

Young children are growing up in what Plowman, Stephen, and McPake (2010b) call an 'e-society' where connectivity to digital technologies is a fundamental feature of daily life. Consequently, the resources available and the ways in which such resources are utilized to support and extend children's play are changing (Marsh et al., 2016).

Miller and Almon (2009) write that there exist wide variations in what is meant by *play*. For the purpose of this chapter, reference to children's play is taken from a cultural–historical perspective (Fleer, 2013). From this viewpoint, through social interaction, children create imaginary scenarios that are based on their everyday encounters (Fleer, 2010). Through the social exchange, children develop new meanings for objects, actions, and their environments. This process therefore becomes a source for learning about the world.

It is also essential to acknowledge that reference to *digital technology* warrants more in-depth considerations due to its complexity. Within the scope of this chapter, a definition for digital technology is taken from Palaiologou's (2016) reference. Digital technology is used as an encompassing term for equipment that contains a computer or microcontroller including laptops or desktop computers, tablets, smartphones, digital cameras, electronic toys, game consoles, and media players.

Included in the reference to digital technologies is the exposure to digital media. As Plowman (2016) states in her article on learning technology at home and in preschool, digital technology and media may be considered inclusive terms. Reference to digital technologies throughout this chapter will thus embody the use of devices such as laptops, tablets, and smartphones as mentioned above in addition to the products or outputs connected to the devices including apps and applications, interactive stories, games, and websites.

With the ubiquitous nature of digital technologies in the everyday lives of young children (Fleer, 2013; Marsh et al., 2016; Stephen & Edwards, 2018), a focus on the implications on children's play is essential (Waller, 2011). Children's play is being reshaped by their exposure and increasing use of digital technologies. Based on what they see in the lives around them, children make reference to or incorporate digital technologies into their play experiences naturally (Arnott, 2017).

The vignette in Appendix A illustrates an example of how one role play scenario evolved with reference to the use of a smartphone.

The vignette demonstrates how children naturally reference the use of digital technologies during their play. In the above instance, the imaginative use of a smartphone enabled the group to develop and extend, what initially began as parallel play, into a more elaborate and complex role play scenario where they were engaged in reciprocal language and a series of exchanges.

In fact, as children grow up in an environment in which digital connectivity is essential to daily life, many young children develop dispositions and competencies through which digital technologies are a context for communicative and social interactions (Waller, 2011). As the lives of young children are immersed in the digital world, it is therefore inevitable that children will incorporate digital technologies in their play.

In an article revisiting the taxonomy of play, Marsh et al. (2016) suggest that a revised definition of play should be offered given the impact of digital technologies on children's play. Fleer (2013) further suggests that the exposure to digital technologies affords a different kind of play. Digital technologies may "serve as catalysts for new forms of play and can have a positive influence on the content of more traditional forms of play" (Fleer, 2017). By traditional types of play, this encompasses role play, gross motor play, and games with rules (Stephen & Edwards, 2018).

In this light, digital play may be the concept used to relate to the accelerating presence of digital technologies in children's play (Edwards, 2018). As used here, the reference to *digital play* means that children are users of digital technologies during their play. In their book about young children's play in the digital age, Stephen and Edwards (2018) write that increasingly the boundaries between traditional and digital play are being blurred as digital technologies are related to the everyday lives of young children. Digital play thus involves children's digital participation in traditional forms of play (Stephen & Edwards, 2018).

We can further align the understanding of digital play to the definition of play using the cultural–historical perspective as discussed above. Where the cultural–historical view sees play occurring within communities and the creation of imaginary scenarios to make meaning (Fleer, 2010), so too can these dimensions be reflected in digital play (Edwards, 2018). As seen in the above role play situation, when children engage in digital play, opportunities to enhance their imaginary situations and to co-construct new understandings are provided.

The below exchange between two three-year-olds is an example of new meaning being constructed through the children's interaction with each other in their imaginary play. They are seated across from each other where one is holding a tablet with a blank screen. What initially begins as an imaginative scenario where the children are pretending to play on the blank screen of a tablet, their play evolves into an experience where one child is challenged by a new perspective.

Constructing meaning through digital play

1 *Lily:* Ok, you hold iPad. I call. [swipes screen]
2 *Molly:* What you do?
3 *Lily:* Call. I call mommy.
4 *Molly:* Why?
5 *Lily:* My mommy working. Mommy is far away. Mommy flying. [holds out arms and imitates the movements of an airplane]
6 *Molly:* But why? Why you call mommy? Why you mommy no home? Where your mommy go?
7 *Lily:* Mommy is working. Mommy go away working.
8 *Molly:* You mommy not home?
9 *Lily:* No, my mommy not home.
10 *Molly:* Oh, but my mommy, my mommy home.
11 *Lily:* My mommy not home. My mommy working. My mommy goes to other place. She go work. Work far away.
12 *Molly:* Mommy work? You?
13 *Lily:* Where your mommy go?
14 *Molly:* My mommy home. My mommy no work.

Through the use of a tablet in their imaginary play situation, what evolved was an elaborate conversation between the two girls where new meanings about their everyday lives were being constructed. The example situates digital play in a social context where the use of a tablet became the tool that facilitated new understandings to develop.

The following discussion draws upon the literature on the nature of play in the digital age to explore the question: *How might digital play be contextualized in children's play at home and in a school environment?*

Contextualizing digital play: At home and at school

As Brito and Dias (2018) indicate, the dynamics between children and digital technologies come about in varied forms and within the two fundamental contexts where children are mainly present, in their home and at school. With the presence of digital technologies in all aspects of the lives of young children, the environments in which children's play take place as well as the way children are playing both at home and in school have been influenced (Fleer, 2013).

Increasingly, it is clear that the nature of play in the digital age is transforming in terms of the resources available for play and the manner in which such resources are utilized during children's play (Marsh et al., 2016). Therefore, as digital technologies are embedded in the everyday play, it is essential that digital play is viewed as a type of play that children inevitably engage in (Stephen & Edwards, 2018).

The following sections explore the opportunities afforded through digital play at home and in a school environment.

Digital play at home

Young children encounter digital technologies first through their immediate environments of their family (Daniela & Rudolfa, 2018; Fleer, 2013; Plowman et al., 2010a; Stephen & Edwards, 2018;). As they observe the adults in their world using digital technologies, children imitate these experiences. As digital technologies are embedded in the ordinary culture of family life, the presence of digital play is thus evident in the children's everyday experiences (Stephen & Plowman, 2014).

The homes in which young children grow up in provide an array of opportunities for learning, creating, and communicating through digital play. Furthermore, with the continual evolution of such tools, restrictions to mobility and access imposed by factors such as size and connectivity are no longer challenges that children face when using digital technologies. With the inevitable use of digital technologies during children's play, not only do young children have *enhanced conditions for playing* but they are also exposed to *differing ways of socializing and interacting* with each other.

Enhanced conditions for play

The range of and possibilities for play activities and experiences may be enhanced through digital play (Stephen & Plowman, 2014). Digital play can be viewed as a way to extend what children experience and are exposed to, provide opportunities to communicate in a multitude of ways and offer additional choices during play through a virtual world.

In his study of children's play using different digital tools and with various programs, Moore (2014) recorded several actions emerging including choice, varying gestural moves, and the use of different tools. **Choice** includes opportunities for children to explore various apps, figuring out what different programs are about. The use of digital technologies can enable children to choose icons to tap on and explore without previous notions about the program and what it may afford. Exploration into various programs also provides opportunities for children to engage with various **gestural actions,** including tapping, sliding, swirling, and the use of a stylus. A combination of gestures may also be required of children. The experience with different gestural moves may be the outcome of sampling with **different virtual tools**. Such tools may include:

- crayons, paintbrushes, pencils, and pens;
- various ways of erasing and deleting;
- possibilities using stamp- and magnet-like objects;
- experimentation with colors, patterns, and backgrounds; and
- exposure to photography and audio.

Through Moore's (2014) study, attention is drawn towards how digital play provides the conditions at home to enhance and extend children's experiences.

In the policy brief put out by the European Early Childhood Education Research Association (Gillen & Arnott, 2018) on the benefits and challenges of digital technologies at home, the research findings on the benefits suggest that children are provided with innovative experiences. When children are engaged in digital play, such experiences allow them to be connected and empowered in new ways. Crucial is the level of collaboration that digital play enables as it can help to generate dialogue and facilitate social interactions. This is discussed in the following section.

Differing ways to socialize and interact

Digital technologies can facilitate opportunities for young children to interact and collaborate with each other, enabling children to become more sociable and cooperative. In an ethnographic study on video gaming, Stevens, Satwicz, and McCarthy (2008) describe the interactions between siblings based on the collection of several vignettes where video game play was a part of the children's lives. The findings from these vignettes illustrated the various levels of collaboration between siblings rather than a solo interaction with the video games played.

The reciprocal interaction provided examples of sociality between siblings as they exchanged strategies on how to play a video game and shared their knowledge where the expert player taught the inexperienced sibling. The study supports the view of digital play from the cultural–historical perspective where:

- the virtual space as in the video game itself provided an imaginary world to its players;
- the social dimension was highlighted through the communication and verbal exchanges that occurred between the siblings whilst playing the video games; and
- the new learning that took place as the more skilled player supported the apprentice.

Table 6.1 Developing the domains of social play through digital play. (Adapted from Broadhead, 2004; Gascoyne, 2012; Quesenberry et al., 2016.)

Associative domain	**Social domain**	**Highly social domain**	**Cooperative domain**
Play involves watching and imitating. An object may be offered. However, the offering may not be accepted and may be ignored. An altercation may be present. **An example:** In this domain, play using digital technologies may involve one sibling watching another sibling. The use of digital technologies may be in isolation. Conflict may arise when determining who uses the tools and when the tools are played with.	• Smiling and laughter may be present. • Play noises and play voices may be used. • Approval is sought, questions are asked, comments are provided. • An offered object is more likely to be accepted. • There may be occasions where altercations arise and children engage in compromise with each other. **An example:** Digital play at this stage involves siblings beginning to interact with each other both through verbal communication and through actions. They may turn take in choosing the tools used and apps played. The interaction can possibly continue after a conflict is resolved.	• Dialogue is evident. • Reciprocal sequences of questions, comments and actions occur. • Objects are offered and accepted within the flow of play. • New ideas begin to evolve through the exchanges. **An example:** Here, siblings engage in digital play that is more sustained. With the increasing level of reciprocal language and action as siblings play a video game to share strategies, the interaction between them involves more opportunities to share, cooperate and socialise throughout the play.	• Dialogue and actions used help sustain the play. • Offered and accepted objects help to extend the play. • New ideas are incorporated into the play as new understandings of the world around them have occurred. **An example:** Digital play at this point provides opportunities for siblings to express their ideas to each other. In playing a strategy video game, the exchange between siblings enable more complex play in the virtual world. They engage in inquiry as questions and problems are posed to them throughout the game. This level of cooperation and interaction between siblings is sustained.

Furthermore, by applying the Social Play Continuum (Broadhead, 2004), the development of social skills and the level of cooperation between siblings can be supported through digital play. Table 6.1 highlights the behaviours that are characteristic of the various domains of social interaction and how the possibilities for siblings to interact and socialize with each other are provided through digital play.

When we focus on the development of social play, digital play can facilitate an array of possibilities in the ways siblings can interact with one another. Digital play provides opportunities for collaboration, communication, and socialization.

Whilst the impact of digital play at home has been highlighted, it is equally as essential to unpack the influences at school. The following discussion will focus on the possible impact of digital play in a school setting.

Digital play in a school environment

While digital technologies may increasingly be accepted as an aspect of early childhood education, Bird and Edwards (2014) emphasize a challenge for early childhood practitioners, posing the dilemma of how best to understand children's development and learning using digital play. With the polarization of the advantages and disadvantages of digital play (Fleer, 2017), it becomes somewhat of a challenge to make appropriate and intentional decisions when it comes to digital play in a school environment.

One of the questions under consideration is how best to balance the quality and quantity of technology use in children's lives (Donohue, 2015; Edwards et al., 2018). As highlighted by Fantozzi et al. (2018) in their article on the integration of digital technologies into children's play, the elements of creation, collaboration, and communication may be applied. Digital play therefore enables:

- New and varied opportunities for playing where apps can be used that are associated with creative activities, including drawing, music making, and video creation (Stephen & Edwards, 2018).
- Interactions that are meaningful where students can collaborate with one another in a two-player game working towards a common goal (Stevens et al., 2008).
- Participation in a social context where a pair of students watch an interactive story together on a tablet and then work together to respond to questions at the end of the story (Quesenberry et al., 2016).

In a whitepaper on the *Six Cs of Deep Learning*, Fullan and Scott (2014) emphasize six qualities critical to the foundation of 21st-century education including:

1 character,
2 citizenship,
3 collaboration,
4 communication,
5 creativity, and
6 critical thinking.

The Six Cs focus on learning that is relevant to the real-world interests and challenges of our students. While it is suggested by the authors that digital technologies can foster the Six Cs, the qualities can effectively be applied through the lens of digital play. Table 6.2 outlines the Six Cs using the original definition of the qualities as well as how the Six Cs can be incorporated into digital play specifically in a school context.

Filtered through the Six Cs of Deep Learning, digital play in a school setting can embody possibilities for collaboration, communication, narration, and transformation. Digital play thus entails "expansion, enrichment, differentiation, individualization and implementation of the globality of goals of the syllabus" (Brito & Dias, 2018, p. 115).

The Six Cs of Deep Learning applied to digital play can also be viewed through the cultural–historical perspective.

Table 6.2 The six Cs in digital play

Six Cs	*The original definition (Fullan & Scott, 2014)*	*The Six Cs through the lens of digital play*	*An example at school (Gerstein, 2013)*
Character	• Developing and practising qualities essential for being personally effective in a complex world	• Digital play can be used to support the development of positive social interactions	Students can listen to social stories online and apply their understanding in a role play situation
Citizenship	• Thinking like global citizens • Considering global issues • Solving complex problems that impact human and environmental sustainability	• Digital play allows opportunities for students to extend their learning beyond the classroom walls, addressing global issues	In a unit of study focusing on the needs of animals, students can build a connection to an artificial life form like caring for a digital pet
Collaboration	• Working interdependently and synergistically • Managing group dynamics • Contributing and learning from others • Making group decisions	• Digital play provides opportunities for students to plan together and collectively contribute towards a project	Students can take pictures of something they built during play. Through the use of an app, they can then narrate and animate their creation together
Communication	• Speaking • Writing • Digital communication	• Digital play diversifies the ways in which students can communicate their ideas	Connections beyond the classroom can be possible where students create and share a video about their learning and post it to their grandparents

(*Continued*)

Six Cs	*The original definition (Fullan & Scott, 2014)*	*The Six Cs through the lens of digital play*	*An example at school (Gerstein, 2013)*
Creativity	• Having an 'entrepreneurial eye' for economic and social opportunities • Asking questions to generate ideas • Pursuing ideas into practice	• Digital play gives new opportunities whereby students are innovative in ways that are not possible without the use of digital technologies	Students can design their own toys. Using a 3D printer, they can create them
Critical thinking	• Evaluating information critically • Observing patterns and connections • Constructing meaningful understanding • Application to the real world	• Digital play enables students to examine, analyse, and evaluate information as well as apply the knowledge into their play	After watching a video on plastics in the ocean, students can apply their learning through role playing the impact plastic has on the environment

- **Social construction:** The qualities of *character, citizenship, collaboration*, and *communication* by definition suggest social interaction between students. There are opportunities for students to co-construct understanding through group work and through a series of exchanges.
- **The imaginary world:** Reflected through the quality of *creativity*, this involves the use of one's imagination. Building on the current knowledge of students, they are then required to imagine something that could be different as they invent and create something new.
- **New meanings:** *Critical thinking* allows students to develop new understandings through a process of asking questions. Students can then analyze and evaluate the information. Eventually, as students synthesize their ideas, new meanings may arise.

The following section will examine some considerations for incorporating digital play into an early childhood classroom.

Digital play in an early childhood setting

According to he Australian Education Union (AEU, 2007) and the National Association for the Education of Young Children (NAEYC, 2009), the internationally accepted definition of early childhood encompasses the range from birth through eight years of age. This definition is applied to the following discussion.

As digital technologies are a prevalent part of the lives and environments in which children play, it is essential to give particular attention to how digital play can be incorporated into an early childhood school setting. Though digital technologies are ever-present in the various aspects of the lives of young children, it is a challenge to develop an approach to digital play. As Verenikina et al. (2016) suggest in their article on applications designed for preschoolers, while children's play is free and self-regulated, digital technologies are designed with rules, guidelines, restrictions, and limitations. This raises the question of how digital play can be incorporated into an early childhood school setting with the idea that play is self-directed.

For young children to engage in authentic forms of digital play, Waller (2011) emphasizes a few considerations including:

- Acknowledgement of the various experiences and competencies in the use of digital technologies. This may result in differing levels of access and understanding about digital technologies.
- Provision of digital play experiences that bridge the interaction with digital technologies at home with that of the school environment and curriculum.

The position statement from the National Association for the Education of Young Children and the Fred Rogers Center (2012) provides guidance for early childhood practitioners about the use of digital technologies in enhancing opportunities for young children's cognitive, social, emotional, and physical development. Key messages from the statement include:

- Special considerations on the use of digital technologies in early childhood education must be given.
- Digital technologies are effective tools to support learning when they are developmentally appropriate and intentionally used.
- Meaningful experiences require practitioners to be knowledgeable and trained about the resources regarding the nature of these digital tools and the implications of their use.

The willingness of adults to facilitate children's digital play is thus a necessity (Paciga et al., 2017). Opportunities to learn through digital play can be made available to children with guidance. This requires practitioners to participate in their students' explorations during digital play (Lindahl & Folkesson, 2012). The practices in the early childhood school setting therefore need to develop in accordance with the children's accounts. "When this is done, by sustaining the norms valuing independent, competent and active children, development of practice is facilitated" (p. 433).

When supported, digital play can be a form of transformative pedagogy (Waller, 2011). As Reggio Children (2018) highlight in their approach to digital environments, the continuous process of experimenting, constructing, and deconstructing provides new ways of exploring and possibilities to emerge. Digital play enables young children to recreate and go beyond what is initially provided. As a result, new and different forms of interaction and play mediated through digital technologies are possible.

Conclusion

This chapter focuses on the impact of the use of digital technologies on the way children play and the resources available for such play. Reference to or the use of digital technologies is infused in the way young children communicate, interact, think, and play.

With the presence of digital technologies in the everyday lives of children, the concept of digital play was used to refer to children's play in the digital age. Grounded in the cultural–historical perspective, digital play was understood as an activity where children socially interact with each other and develop imaginary scenarios that reflect their experiences. Through this exchange, new meanings are created.

A discussion of digital play in the contexts of the home and school was included. At home, digital play provides opportunities for enhanced play environments, offering new possibilities for play scenarios to evolve. Digital play also enables siblings to interact with one another in different ways. The Social Play Continuum (Broadhead, 2004) was used to look at how the interaction and socialization of siblings can be developed through digital play. At school, digital play provides students with opportunities to develop the Six Cs of Deep Learning (Fullan & Scott, 2014). These qualities include building character, developing citizenship, opportunities for collaboration, diversity in ways to communicate, additional avenues to use one's creativity, and chances to develop critical thinking.

The following are some considerations for incorporating digital play specifically to an early childhood school context. To develop practices surrounding digital play, fundamental is the role of the adult in (1) understanding the previous experiences of young children with regard to digital technologies and (2) facilitating digital play by providing the environment that is both intentional and developmentally appropriate.

In conclusion, while digital play provides enhanced ways for children to play, as Donohue and Schomburg state (2017), when it comes to understanding and developing digital play, "there is still much to learn" (p. 77).

References

Arnott, L. (2017). Framing technological experiences in the early years. In L. Arnott (Ed.), *Digital technologies and learning in the early years* (VitalSource ed.). Retrieved from https://bookshelf.vitalsource.com/#/books/9781526414472/cfi/6/24!/4/2/4/2@0:0

Australian Education Union. (2007). Early childhood education policy 2007. Southbank: Australian Education Union.

Bird, J., & Edwards, S. (2014). Observing and assessing young children's digital play in the early years: Using the digital play framework. *Journal of Early Childhood Research, 15*(2), 158–173.

Brito, R., & Dias, P. (2018). Digital technologies in kindergarten: Paths of kindergarten teachers and potentialities for children. In L. Daniela & M. Lytras (Eds.), *Learning strategies and constructionism in modern education settings* (pp. 114–130). Hershey, PA: IGI Global.

Broadhead, P. (2004). *Early years play and learning: Developing social skills and cooperation.* New York: Routledge.

Daniela, L., & Rudolfa, A. (2018). The role of parents for developing digital literacy of 0–5 year olds. In L. Daniela & M. D. Lytras (Eds.), *Learning strategies and constructionism in modern education settings* (pp. 104–113). Hershey, PA: IGI Global.

Donohue, C. (2015). Technology and digital media as tools for teaching and learning in the digital age. In C. Donohue (Ed.), *Technology and digital media in the early years: Tools for teaching and learning* (pp. 21–35). New York: Routledge.

Donohue, C., & Schomburg, R. (2017). Technology and interactive media in early childhood programs: What we've learned from five years of research, policy, and practice. *Young Children, 72*(4), 72–78.

Edwards, S. (2018). Digital play. In *Encyclopedia on early childhood development.* Retrieved from http://www.child-encyclopedia.com/play-based-learning/according-experts/digital-play

Edwards, S., Straker, L., & Oakey, H. (2018). *Early childhood Australia: Statement on young children and digital technologies.* Canberra: Early Childhood Australia.

Fantozzi, V. B., Johnson, C. P., & Scherfen, A. (2018). Play and technology: An important intersection for developing literacy. *Young Children, 73*(2), 88–93.

Fleer, M. (2010). *Early learning and development: Cultural-historical concepts in play.* Melbourne: Cambridge University Press.

Fleer, M. (2013). *Play in the early years.* New York: Cambridge University Press.

Fleer, M. (2017). Digital pedagogy: How teachers support digital play in the early years. In L. Arnott (Ed.), *Digital technologies and learning in the early years* [VitalSource ed.]. Retrieved from https://bookshelf.vitalsource.com/#/books/9781526414472/cfi/6/24!/4/2/4/2@0:0

Fullan, M., & Scott, G. (2014). *New pedagogies for deep learning whitepaper: Education PLUS.* Seattle, Washington: Collaborative Impact SPC.

Gascoyne, S. (2012). *Treasure baskets and beyond: Realizing the potential of sensory-rich play.* Maidenhead: Open University Press.

Gerstein, J. (2013). *Let children's play (with technology) be their work in education.* Retrieved from https://usergeneratededucation.wordpress.com/2013/01/12/let-childrens-play-with-technology-be-their-work-in-education/

Gillen, J., & Arnott, L. (2018). Digital literacy and young children: Towards better understandings of the benefits and challenges of digital technologies in homes and early years settings. Retrieved from https://www.researchgate.net/profile/Lorna_Arnott/publication/327594155_Digital_Literacy_and_Young_Children_Towards_Better_Understandings_of_the_Benefits_and_Challenges_of_Digital_Technologies_in_Homes_and_Early_Years_Settings/links/5b98e41945851531057ef9b1/Digital-Literacy-and-Young-Children-Towards-Better-Understandings-of-the-Benefits-and-Challenges-of-Digital-Technologies-in-Homes-and-Early-Years-Settings.pdf?origin=publication_detail

Lindahl, M. G., & Folkesson, A. M. (2012). ICT in preschool: Friend or foe? The significance of norms in a changing practice. *International Journal of Early Years Education, 20*(4), 422–436.

Marsh, J., Plowman, L., Yamada-Rice, D., Bishop, J., & Scott, F. (2016). Digital play: A new classification. *Early Years, 36*(3), 242–253.

Miller, E., & Almon, J. (2009). Crisis in the kindergarten: Why children need to play in school. College Park, MD: Alliance for Childhood.

Moore, H. L. C. (2014). *Young children's play using digital touch-screen tablets.* (Unpublished PhD thesis). The University of Texas at Austin. Austin, Texas.

National Association for the Education of Young Children. (2009). Developmentally appropriate practice in early childhood programs serving children from birth through age 8. Washington, DC: Author.

National Association for the Education of Young Children and the Fred Rogers Center for Early Learning and Children's Media. (2012). *Technology and interactive media as tools in early childhood programs serving children from birth through age 8* [Position statement]. Retrieved from https://www.naeyc.org/sites/default/files/globally-shared/downloads/PDFs/resources/topics/PS_technology_WEB.pdf

Paciga, K. A., Donohue, C. D., Struble Myers, K., Fernades, R., & Li, J. (2017, May). *Carrying Fred Rogers' message forward in the digital age.* Fred forward symposium proceedings. Latrobe: Fred Rogers Center for Early Learning and Children's Media at Saint Vincent College.

Palaiologou, I. (2016). Children under five and digital technologies: Implications for early years pedagogy. *European Early Childhood Education Research Journal, 24*(1), 5–24.

Plowman, L. (2016). Learning technology at home and preschool. In N. Rushby & D. Surry (Eds.), *Wiley handbook of learning technology* (pp. 96–112). Chichester: Wiley.

Plowman L., Stephen C., & McPake, J. (2010a). *Growing up with technology: Young children learning in a digital world.* London: Routledge.

Plowman, L., Stephen, C., & McPake, J. (2010b). Supporting young children's learning with technology at home and in preschool. *Research Papers in Education, 25*(1), 93–113.

Quesenberry, A. C., Mustian, A. L., & Clark-Bischke, C. (2016). Turning in: Strategies for incorporating technology into social skills instruction in preschool and kindergarten. *Young Children, 71*(1), 74–80.

Reggio Children. (2018). *Digital environments: Technological invitations and luminous ideas.* Reggio Emilia: Reggio Children.

Stephen C., & Edwards, S. (2018). *Young children playing and learning in a digital age: A cultural and critical perspective.* London: Routledge.
Stephen, C., & Plowman, L. (2014). Digital play. In L. Brooker, M. Blaise, & S. Edwards (Eds.), *Sage handbook of play and learning* (pp. 330–341). London: SAGE Publications.
Stevens, R., Satwicz, T., & McCarthy, L. (2008). In-game, in-room, in-world: Reconnecting video game play to the rest of the kids' lives. In K. Salen (Ed.), *The ecology of games: Connecting youth, games and learning: The John and Catherine MacArthur foundation series on digital media and learning* (pp. 41–66). Cambridge, MA: MIT Press.
Verenikina, I., Kervin, L., Rivera. M., & Lidbetter, A. (2016). Digital play: Exploring young children's perspectives on applications designed for preschoolers. *Global Studies of Childhood, 6*(4), 1–12.
Waller, T. (2011). Digital play in the classroom: A twenty-first century pedagogy? In S. Rogers (Ed.), *Rethinking play and pedagogy in early childhood education: Concepts, contexts and cultures* (pp. 139–151). Milton Park: Routledge.

Chapter 7

Materialities in innovative education

Focus on small Italian schools

Giuseppina Rita Jose Mangione[1] *and Samuele Calzone*

Cultural framework of educational materiality in small schools

The materiality of learning represents a specific and outlined trajectory of pedagogy, today at the centre of the ongoing educational debate on how digital technology is changing the way of schooling (Fenwick et al., 2015). The phrase *educational materiality* (Barone, 1997; Mantegazza, 2001; Massa, 1997) puts forward a theoretical construct that looks at practices as a set of *concrete determinations* (Ferrante, 2016) that make education possible.

Such an approach shifts the focus towards both educational action and multiple interrelations that occur between subjects and *objects of a didactic culture* (paper documents, diaries, desks, chairs, work spaces, furnishings, community, etc.) existing within educational contexts, and often subject to *re-mediation* by new technologies. Thus, material components previously observed as "peripheral" or "exotic" elements (Landriscina & Viteritti, 2016) or, at most, from the point of view of an archaeology of objects (Lawn & Grosvenor, 2005), prove pivotal for the social and educational spaces of school, which is often regarded as a missing and neglected mass (Latour, 2006). Massa's (1997) and Pontecorvo's (2000) works had already attributed an important role to objects, spaces, and devices in learning processes. Scholastic innovation in Italy, together with scientific reasoning on it, has always developed from a reflection perspective on materiality (think of the works by Montessori, Lodi, Malaguzzi), yet, it has never been truly capable of leading change in educational practice.

The relationship between postmodernism and education, as well as the impact that technologies have been having on revisiting "scholastic space and time" and on identifying "educational events" (Landriscina & Viteritti, 2016; Masschelein & Simons, 2013) have led to overcome the idea of school as an isolated *oasis* or *forme scolaire* (Postman, 1979). Hence, also the challenges of change have been reconsidered as for: Integration of new technologies in educational practices (Nespor, 2011), processes of teaching enhancement (Mulcahy, 2011), study of educational policies (Gorur, 2011), curriculum implementations (Edwards & Carmichael, 2012), and study of educational spaces (McGregor, 2004; Paechter, 2004).

Consequently, materiality becomes an interpretative category of educational innovation (Fenwick & Landri, 2012) (Fenwick et al., 2012). In this work, it is a key concept that allows us to decode the practices proposed by Italian small schools, and to outline a pedagogical object to be studied theoretically and empirically.

Small school is classified as a non-standard educational context, peculiarly *fragile* because of the *inequality* (Bouck, 2004; Harrison & Busher, 1995) often attributed to geographically isolated areas and at a high risk of early school leaving. The debate on *small schools* divides the international scientific community in the field of teaching and learning, highlighting two opposing viewpoints. On the one hand, is the idea of small school as a place implementing a disadvantaged teaching, and unable to provide pupils with equal educational opportunities, especially for all those who aspire to achieve high educational levels (Bouck, 2004; Harrison & Busher, 1995). On the other hand, a different vision that interprets small school as a place for innovation and pedagogical experimentation (Kvalsund & Hargreaves, 2009; Strike, 2008), and that converges on the hypothesis that adoption and experimentation of sustainable teaching strategies and practices are essential grounds for change, making *small schools* also *good schools* (Mangione & Calzone, 2019).

The work aims to explore how digital-mediated materials support the teaching action of schools adhering to the Italian Small Schools Movement,[2] a national network promoted by Indire, with more than 200 members and a manifesto[3] that traces trajectories of change, such as memory community and learning quality, multi-grade classes experience – a resource, not a limit – technologies and social inclusion.

To do so, we have tried to identify the possible *dimensions of materiality*, here understood as an intrinsic component of the *social sphere*, and as a network inside and outside the classroom (Landriscina & Viteritti, 2016).

Forms of *remediated materialities* to support vertical curriculum and development of competencies. With reference to a popular pedagogy initiated by Freinet and adopted in Italy by the Educational Cooperation Movement, an alternative model to transmissive school has emerged. This model either changes the traditional classroom's spatial layout, being reorganized into corners for different activities (reading, printing, laboratory, atelier) or allows to set up alternative teaching–learning (inter-school correspondence, self-correcting diaries, class library, free text active drafting), which is going to break rigid class block patterns and to implement a dynamic and flexible group work. In such a context, a *Cooperative classroom* becomes a functional instrument for multi-class school – where different grades are present in the same classroom – and englobes a model of shared learning management that rests on a pedagogical, social, and value-driven approach intended for all school levels. The cooperative classroom passes through and includes all school adjacent "physical" spaces. In this sense, *the courtyard* (Patriarca, 2018) may not only help develop relational and mediation skills, but also connect corporeality to learning basic subjects like mathematics (a courtyard, for instance, may be used for creating maths games, explaining concepts such as the ratio between the time it takes to reach a goal and the time it takes to count in the hide-and-seek game, or teaching measurement units). Similarly, *the walls* (Turrisi, 2018) may be seen as materialities to be used for representing narration moments and would help retrace the steps took to reach the given results, and highlight references to school model, values, and educational guidance (based on identity recognition, inclusion, etc.). The walls – as *materiality* that reclaims other materialities, such as wall charts and virtual globes – would encourage pupils develop their geographical skills. Objects inspire a kind of teaching that guides autonomy and awareness or, as Thaler argues, they contribute to that gentle push or nudge framed on the ethics of choice architecture (Thaler & Sunstein, 2014).

Examples of how to attribute a didactic function to furnishings are "colour spaces": That is to say that different settings support diverse learning strategies aimed at bridging the rift between theory and practice (Marcarini, 2018). In this concept, study spaces and facilities, as well as diffused libraries and exhibitions are co-designed and made accessible through shared rules to develop pupils' environmental awareness (Martin, 2002).

Forms of *educating city* for widespread and inclusive learning. Moving from schools typically based on subjects, schedules, desks, and walls, to schools aimed at widespread education. For this purpose, the urban fabric is revised to be committed to an educating function, as well as the places of interest (Campagnoli, 2018) become relevant aspects in the educating city. The educating city is the result of evolution as initiated by avant-garde school architecture that, in the context of educational materiality, leads to think of *educational portals* to be implemented over urban areas, intended to encourage the creation of classrooms/laboratories outside the school walls. On this issue, it seems necessary to resume a collective urban narrative going beyond the exasperated functionalism of industrial revolution and *zoning*, and to integrate all efforts already made for restoring school buildings so to expand the concept of the *living room* from the building to the city district. The new spaces will be multifaceted and be able to complement the classroom: A square, a shop, a theatre, a local museum, sons of a pedagogy that gives a global look at the community. Community and landscape are seen as third educators.

Forms of *educational centrifugal materialities* for strengthening competencies. These are actions that develop in *emergency* cases, but which, paradoxically, make their nuclear and familistic traits an educational extension of the school. An example of this is the *home-schooling movement* widespread in the United States, which considers home as a place of education, or also *shadow education*, in Japan and other Southeast Asia countries with a strong industrial connotation, where families take charge of private supplementary tutoring not only to assist their children with after-school program but to expand and compensate for what the school alone may not be able to offer (Baldascino, 2018). It is not a matter of supporting forms of de-schooling, but as in the case of small schools, to ask for a family support in situations of hardship and isolation (let us think, for instance, of seismic areas that, after local schools are destroyed by earthquake, force pupils to move to the nearest towns' institutes) or when there are no appropriate didactic resources for personalized teaching–learning (such as support for Special Educational Needs or gifted children).

Forms of *meccagogia* for social and cultural inclusion (Baldascino, 2018). That is, a *machine friendly pedagogy* that uses virtual and intelligent educational systems to improve and manage classroom situations. Classrooms today are *connected* and *technology-enhanced*, and accommodate scholastic and extra-curricular activities (Mangione et al., 2017). On the one hand, they include a framework for Future Lab classroom and classrooms 3.0, and are endowed with *special furnishings* for group working, portable devices recharging, and innovative *applications* for classes 3.0 and all devices, including cloud apps like Opedia – an interaction system between blackboard and tablet, allowing engaging lessons and enabling pupils to take notes, store them on their own cloud profile, and access them at home, along with other contents used in the classroom. On the other hand, they promote initiatives, experiences, and paths for teachers, school staff, and pupils, such as *fab labs*, educational robotics and coding competitions, hackathons, flash mobs, digital storytelling, teachers' workshops, and activities for families, also documented through live stories on social networks. The classroom extends and connects educational experiences to places, and in doing this it enables knowledge activation. Possible places can be public squares equipped with digital installations intended to host school robotics competitions, as well as immersive spaces, fab labs, digital treasure hunts (Digital Circus), Italian schools' digital good practices (Future Zone). City places can also host innovative educational workshops for students (Students Matter), areas dedicated to competitive paths on innovative teaching methods, like *Civic Arena*, where teams of pupils, winning local hackathons, challenge each other in the creation of digital solutions to be applied in the future in sectors of innovation and sustainability, or *Womest*, dedicated to teams of pupil girls, winning local STEM (Science, Technology, Engineering, Mathematics) Labs, who will play the *Future City Lab* final.

After identifying the materiality dimensions, it has been possible to analyse the innovation introduced by the Italian small schools adhering to the project Small Schools Movement[4] through projects implemented in response to *Public notices* issued under the 2014–2020 PON for Schools[5] in favour of an inclusive school.

The analysis was carried out based on the following question: "*What are the emerging forms of educational materiality*"? In the following sections, both quantitative and qualitative aspects are discussed in more detail. Firstly, we will set forth a picture of small schools involved in all PON notices, to focus, afterwards, on the responses to notices solely concerning "heritage", "dispersion", and "basic competencies". Lastly, we will read – through *comprehensive approaches* looking at materiality as a key tool of change – some schools' projects also selected according to their capacity to revise environments and workspaces.

The opportunities of the 2014–2020 National Operational Programme

The National Operational Programme (PON) for "skills and learning environments", active for the 2014–2020 programming period, provides support for the national education and training system, and more generally the implementation of EU strategies. Conceived as a vector of sustainable and inclusive growth that is able to steer Europe out of the latest crisis, the programme promotes raising and enhancing the overall competencies of both young people and adults (smart growth); it is in this sense that

> it intervenes to counteract school dropout, to improve both the quality of education system and school attractiveness, by strengthening learning environments, promoting dissemination of specific competence, and supporting a process of school innovation and digitalisation.[6]

The programme, which is divided into less developed, in transition, and more developed regions,[7] acts on the whole national territory (the schools involved are around 8.300) from a system perspective. As a matter of fact, to support these interventions, which contribute to achieving a *school open to the territory*, transforming it into a civic centre,[8] it mainly makes use of two Funds: European Social Fund (ESF) that deals with training initiatives aimed at students, teachers and local communities (raising of competencies) and European Regional Development Fund (ERDF), which is intended to improve attractiveness of schools, increasing technological equipment and interventions on school buildings. These two Funds complement each other for the creation of measures to combat school dropout: Accordingly transforming the school into *open school*, a social gathering place for local communities, means creating and sustaining ways to promote initiatives for preventing youth hardship and disadvantage, among whose many effects that have been monitoring today, are precisely school dropout and loss of skills triggering the risk of social exclusion. Some authors, in fact, define social exclusion as

> a concoction (or blend) of multidimensional and mutually reinforcing processes of deprivation, associated with progressive dissociation from social milieu, resulting in the isolation of individuals and groups from the mainstream of opportunities society has to offer;
>
> (Berghman & Salais, 2001)

hence, developing skills and expanding knowledge becomes an important contribution of the Programme to smart, inclusive and sustainable growth (EU 2020 Strategy).

With a view to achieving a gradual improvement, the Programme pays particular attention to schools whose strong fragility may depend on some variables, mainly geographical and socio-economic. The Italian territory also consists of isolated, cut off areas, both characterized by precarious economic conditions and having difficulty connecting with peninsula's backbones of development. Schools in these areas are widespread and are identified with the name of Small Schools (Piccole Scuole), generally attended by a few pupils that are also grouped in multiple classes; these schools represent an educational and cultural defence against the phenomenon of depopulation.

Funds monitoring and guidance, in particular for disadvantaged geographical areas presenting the highest level of fragility, is allowed through a system for online *governance* (GPU), developed by INDIRE.[9] The Italian 2014–2020 PON 'For School' Management Authority[10] contributes to the implementation of activities of analysis, planning, activation and control aimed at improving the Programme management at all levels. Through pre-set formats, this system grants a structured collection of data and guides the involved subjects towards completion of the procedures required by the Programme, also enabling high levels of transparency, communication and sharing. In other words, all schools intending to participate in the Programme initiatives respond to a *Notice* available in the system by submitting a project that is organized on the basis of either training or local area's needs; if authorized, the project is managed in the system: All procedures, products, modifications must be documented directly online. This system allows for a collection of digital learning/teaching units (over 250.000 products) that feed a school data bank gathering the materials produced and that are available to students.

In recent years, the Programme has launched initiatives for school improvement, promoting an increase in the technological equipment, and supporting specific training activities. In the first case (ERDF Fund), the notices examined for this work are specified as follows. The first one (Notice No. 9035/2015) mainly allowed to "build infrastructure and access points to LAN / WLAN" for schools of the first cycle of education (3.514 schools); while, for the second cycle, it mostly concerned "expansion or upgrading of infrastructure and access points to LAN / WLAN and enhancement of cabling installation" (1.177 schools). The second notice (Notice No. 12810/2015), both for first and second cycle, focused on the design of "technology-enhanced classrooms" (digital environment), defined as

> traditional classrooms endowed with facilities for collective and individual use of web and web-content, and interaction of different aggregations within learning groups, via wired or wireless connection, towards a daily integration of digital devices in teaching.[11]

These notices respond to the need to accept change in the educational paradigm that

> heads in the direction of a definitive overcoming of transmissive teaching, and transforms the 'giving lessons' scenario, being enriched and enhanced with new and multiple 'educational' laboratories for achievement of knowledge and competencies.[12]

This renewal requires rethinking educational and training spaces, promoting school that complements to local community and national productive system, in which

> organization and delivery of services, educational and training activities be intended to ensure everyone had full access, good inclusion practice, and high levels of competency.[13]

As for the training initiatives (ESF Fund), the following notices are examined. The first one (Notice No. 10862/2016) is aimed at "projects of social inclusion and fight against disadvantage to guarantee schools opening beyond school time, especially in geographically isolated or cut off areas". The second notice (Notice No. 1953/2017) is aimed at state schools for innovative strengthening of basic competencies (strengthening linguistic, expressive, relational and creative learning, as well as developing competencies in the mother tongue, in the foreign language, in mathematics and sciences). The last one (Notice No. 4427/2017) is aimed at state schools for strengthening education in cultural, artistic and landscape heritage (improving cross-cutting competencies, especially skills aimed at spreading business culture).

These notices stress the importance of developing competencies (key and cross-cutting ones) to improve pupil's involvement and, more generally, school welfare; Notice No. 10862, in particular, has

> as its primary objective to rebalance and compensate for socio-economic disadvantage in particularly disadvantaged, isolated or cut off areas, through actions targeted to groups of pupils with specific difficulties and needs, and at a higher risk of drop-out, but also involving other subjects: local government public bodies, associations, foundations, professionals.[14]

Schools reported high rates of participation in the initiatives and bore witness to the interest and importance such actions have for the Italian education system. In fact, more than 660.000 students got involved[15] in the actions promoted by Notice No. 10862/2016, for a total of 4.536 schools, with less developed regions (Basilicata, Calabria, Campania, Puglia and Sicilia) involving the largest number of pupils, namely 390.053, i.e., over 245.000 pupils more than those from more developed regions.[16] Notice No. 1953/2017 concerns the development of key competencies, for a total of 5.228 schools; the higher number of pupils involved is from less developed regions, i.e., over 304.000 more, i.e. over 219.000 pupils from more developed. As for Notice No. 4427/2017 (2.845 schools involved), the above trend is reversed: More developed regions had the greatest impact in terms of participation with 34.000 pupils more than in less developed regions (32.000 pupils) (Table 7.1).

These considerations are at root of the next paragraph that will explore selected projects and Small Schools involved into the initiatives financed under the above mentioned ESF Fund Notices on school early leaving (Notice No. 10862), basic skills (Notice No. 1953/2017) and cultural, artistic and landscape heritage (Notice No. 4427/2017).

Table 7.1 Participation of Italian schools, divided by region in transition, less developed and highly developed, to PON notices

Region	*No. of schools notice no. 10862*	*Enrolled pupils*	*No. of schools notice no. 1953*	*Enrolled pupils*	*No. of schools notice no. 4427*	*Enrolled pupils*
Abruzzo	94	14.088	106	8.371	73	954
Molise	35	4.645	33	3.059	26	591
Sardegna	106	9.128	90	5.511	44	374
In transition (total)	**235**	**27.861**	**229**	**16.941**	**143**	**1.919**
Basilicata	60	7.167	74	4.268	43	496
Calabria	323	49.355	316	27.976	185	2.780
Campania	844	140.353	855	117.227	482	13.485
Puglia	552	92.040	573	77.253	324	7.671
Sicilia	683	101.138	717	77.549	428	8.384
Less developed (total)	**2.462**	**390.053**	**2.535**	**304.273**	**1.462**	**32.816**
Emilia Romagna	205	30.327	248	24.752	127	3.508
Friuli Venezia Giulia	66	6.906	103	7.859	56	1.419
Lazio	333	44.542	420	31.331	239	3.786
Liguria	67	8.331	97	7.967	51	1.362
Lombardia	378	47.850	499	46.204	217	5.990
Marche	110	15.775	136	11.109	70	2.652
Piemonte	224	29.120	321	30.527	151	5.997
Provincia Aut. Trento	0	0	9	330	0	0
Toscana	220	29.034	286	26.185	157	3.835
Umbria	81	13.072	102	13.020	56	2.402
Veneto	155	20.399	243	20.472	116	3.310
More developed (total)	**1.839**	**245.356**	**2.464**	**219.756**	**1.240**	**34.261**
Total national basis	**4.536**	**663.270**	**5.228**	**540.970**	**2.845**	**68.996**

Small schools profile for an educational materiality: A quantitative analysis

As a result of a number of reforms in education and training, important national and community initiatives have been launched in recent years to encourage the concept of an *inclusive school*. In particular, small schools, which represent a frontier reality that is yet to be explored and deepened, are the content of a specific area of study: Italian Law No. 111 of 2011 (and subsequent amendments) in fact, grants autonomy to schools enrolling up to 600 pupils, reduced to 400 for schools located on the islands, in mountain areas or contexts characterized by linguistic specificity or fragility. Hence, small school as a field of experimentation to investigate educational materiality is an interpretative category that we may decline according to dimensions opportunely identified in the first section.

Before proceeding further with our analysis, which will be explored in the next paragraph, let us outline a small schools profile according to the following characteristics:

- a at least 75% of school complexes are "small complex", i.e. they have less than 200 pupils per premise
- b all school facilities are located in "inland areas".[17]

Based on this first analysis, we have identified 611 small schools, 70 of which adhere to the Small Schools Movement,[18] and are divided according to the territorial distribution given below.

Although small schools are mostly distributed in the southern Italian regions (Calabria, Campania, Sicilia), in the Calabria region there are no schools adhering to the movement, which instead has a higher participation in Campania, Abruzzo, Basilicata, Toscana, and Sardegna. The movement's paramount objective is to create a network of schools that offer solutions to keep school population in the area, facilitating lifelong learning also through the potential of digital technologies, as to promote shared and distance learning, and an extended learning environment that allows to enhance pupil's skills and provide them with opportunities to continue their education (Figure 7.1).

To further deepen the identified small schools profile, we have analysed their participation in the PON notices.

With regard to the 611 small schools, the largest number of schools participated in the ERDF Programme through the notice aimed at digital environments (443 schools), while fewer than 200 schools were interested in the notice concerning "cultural heritage" (Table 7.2).

The 70 schools adhering to the movement, compared to all of identified small schools (611), carried out projects aimed at improving the infrastructure and technological equipment with a slightly higher percentage, while these participated in the ESF Programme Fund initiatives to a lesser extent (even by 14 percentage points) (Figure 7.2).

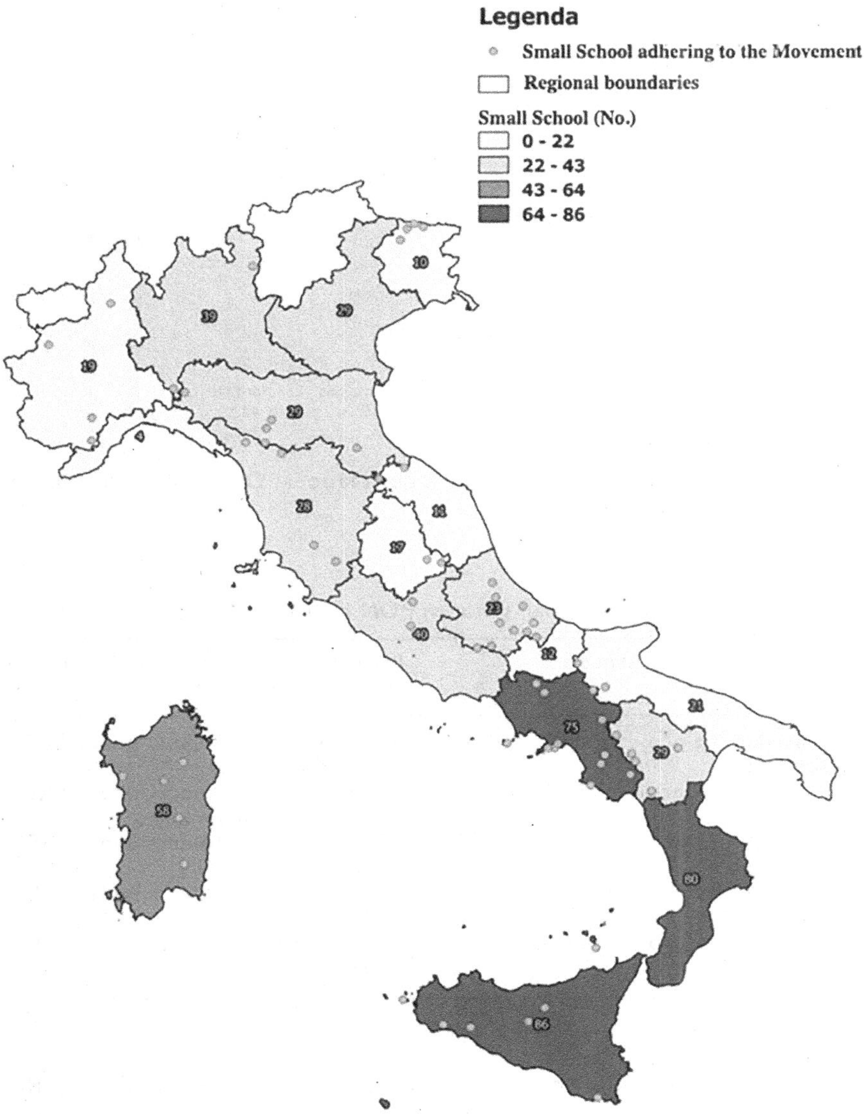

Figure 7.1 Cartogram of Italy's distribution of small schools and small schools movement.

What accounts for a lower participation in the European Social Fund (ESF) Notices is possibly the number of other initiatives carried out by schools adhering to the movement: These are, in fact, schools that are generally active in their area through a variety of projects, both national and regional, involving

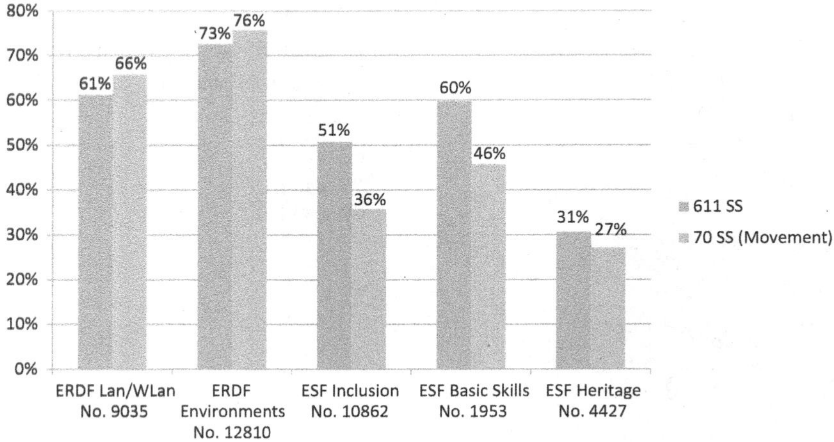

Figure 7.2 Distribution of small schools per National Operational Programme (PON) notices.

Table 7.2 Participation of small schools, PON notices

Participation of small schools	*Participation of small schools adhering to the movement*
374 carried out projects under ERDF "Lan/WLan"	46 carried out projects under ERDF "Lan/WLan"
443 carried out projects under ERDF "Digital Environments"	53 carried out projects under ERDF "Digital Environments"
310 carried out projects under ESF "Inclusion"	25 carried out projects under ESF "Inclusion"
365 carried out projects under ESF "Basic Skills"	32 carried out projects under ESF "Basic Skills"
187 carried out projects under ESF "Heritage"	19 carried out projects under ESF "Heritage"

the whole local community. These schools show a distributed leadership that improves *governance* actions: School resources and staff are organized with a systemic view to enhance the role of the school in the territorial context.

Some of the schools involved in the movement (seven schools), subject of a supplementary qualitative analysis meant to investigate the possible dimensions of educational materiality, were able to participate in all programme initiatives indicated earlier in the paragraph (ERDF for LAN / WLAN, ERDF for Digital Environments, ESF for Dispersion, ESF for Basic Skills, ESF for Heritage). With the exception of Edolo (Brescia) and Paganico (Grosseto),

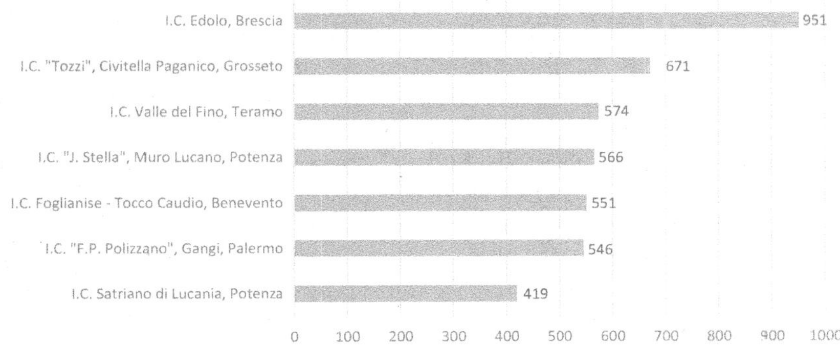

Figure 7.3 Number of pupils per small school.

which include many school complexes, 12 and 15, respectively, the number of pupils in other schools are between 400 and 500 (Figure 7.3).

As regards supplies and equipment purchased for the creation of digital environments, the seven schools involved in the movement showed differences. Alongside their decision to adopt "devices for dematerialization and digitization of documents and processes" or "Software for storage and production of integrative multimedia contents", which accompanies all the identified schools, some of them focused on the development of inclusive teaching that adopts information and communications technology for disability: For example, I.C. Polizzano di Gangi (Sicily), purchased "Hardware aids for users with disabilities to be able to utilize technological devices" and "3D printers" to promote projects for the participation of disabled students. Some schools, on the other hand, paid particular attention to music and its connection to education. Our culture, in fact, gives music a pedagogical role intended both as making music with an instrument or as singing, but also listening to music, then focusing on the use of language and words. Since the time of Plato's philosophy (in Protagoras dialogue), in fact, music has been regarded as a form of control over one's passions, self-control, and inner balance that educational renewal – smart pedagogy – is called, nowadays, to intercept and promote. I.C. Joseph Stella di Muro Lucano (Basilicata), for example, fostering an inclusive culture, dealt with digital music by purchasing digital piano, electronic guitar, drums and keyboard, and software for making music in the classroom, as supplies under ERDF Notice for 'Digital environments'.

As for actions to combat early school leaving, all selected small schools (seven schools) achieved good results: They carried out training courses to improve school welfare, encouraging greater collaboration and participation. The courses were organized according to three thematic axes: Sport, theatre, strengthening of linguistic and scientific skills. On average, about one-third of the pupils were involved in these educational initiatives, with the exception of those of Satriano di Lucania, of which 54% pupils (225) participated.

It should also be noted that before and after taking part in a course financed under notice No.10862, all pupils' syllabus grades in subjects such as *Italian*, *foreign languages*, *mathematics*, and *sciences*, were registered in the online governance system (GPU). Grades achieved after participating in this initiative of the programme are generally higher[19]: In some cases, failing grades are recovered, and excellence cases are strengthened.

Participation in the programme has contributed, in the selected schools, to a positive school climate, which, as numerous studies testify (for example Mitchell, 2008; Wentzel, 1997), underpins any learning improvement, in terms of pupil's successful schooling and well-being. Since the early years of the 20th century, John Dewey's pedagogical proposal (Dewey, 1970), relaunched and deepened in the 1990s and taken up in recent times, argues the importance of the relationship between the individual and the environment: A quality-based educational environment consists of a system of positive interactions in support of pupils.

The above schools accepted the opportunities offered by the programme to create an educational environment fostering scholastic well-being. In the following paragraph, upon these considerations, their projects will be analysed to intercept how educational materialities may be seen as helpful tools to enhance school community and, more generally, the territorial community where small schools operate.

Qualitative analysis of projects

The Italian scientific community connected to the subject of teaching and learning in small schools located in isolated areas with high risk of depopulation, converges on the hypothesis that the diffusion and experimentation of sustainable teaching strategies and practices are grounds for change. Small schools can be an opportunity to experiment with new forms of educational "materiality" and to innovate curriculum and teaching methods. The Manifesto of Small Schools Movement[20] highlights that *time and space* of a small school and its relationship with local community is able to offer unique opportunities to move towards the school of the future. Large classes and insufficient number of buildings without surrounding spaces often represent an obstacle to the creation of innovative learning environments. In schools situated in rural areas and on islands, which deal with small numbers, surprisingly potential advantageous conditions are determined, to transform educational environments into places of inclusion and experimentation.

Privileging an integrated survey approach, this part of the work presents a description of projects carried out by seven small schools selected through a quantitative analysis for their capacity to systematize opportunities offered by the 2014–20120 PON "For School", according to an innovative vision or need to respond to a problem. The description is accompanied by qualitative–interpretative comments that take into account the multidimensionality of the context being investigated and help understand transformative and innovative processes generated by the projects in those areas. Innovative experiences documenting gives us a chance to understand the impact that projects, if consistent with each other, may represent in communities where small schools intervene. The projects are investigated through an inductive approach underlying interpretative processes of qualitative analysis, and taking into account the materiality dimensions identified in the section 'The cultural framework of educational materiality in small schools'. These dimensions should be intended as cores of meaning or codes, resulting from an early process of inference, referable to the context and the scientific literature on educational materiality.

In a subsequent phase, possible types of relationships between emerged cores/codes (Johnson & Onwuegbuzie, 2004) and school projects (Semeraro, 2014) are identified to retrieve insight trajectories about experiences with respect to didactical innovation.

Such a qualitative reading, thus, allows us to pay attention to features such as *complexity* (multidimensionality of small schools educational experience or in isolated and marginal areas), *contextuality* (the proposed experiences must be interpreted based on the small school's situational context) and *processuality* (Small Schools' projects and innovative experiences are also strongly dependent on a time dimension, characterized by parallel cultural and political actions aimed at supporting the development of educational actions for small schools).[21]

Our reading has been guided by in-depth categories, built on a transversal reading of projects per school, which allowed us to comprehend their design consistency with specific vision or trajectories of educational innovation belonging to each school.

Schools' projects for strengthening basic competencies, inclusion, and cultural heritage education

The 15 school complexes of **I.C. Tozzi – Civitella Paganico (GR)** are located in a portion of territory of about 500 sq. km. Primary and secondary schools suffer from marginality, despite their cultural, artistic, and naturalistic potential. The environment is rural and its inhabitants are mostly engaged in primary sector activity. Many students live in isolated farms, poorly connected to road network, spending most of their leisure time by themselves. On average, families' economic and cultural status is low and immigrants' percentage is overall 26% in the area served by the Institute, with peaks of 46%. Most immigrant families are not integrated into the social fabric.

The analysis of projects brings out an enhancement of *school as an active community, open to the territory, and able to develop and increase interaction with families and local community, including third-sector organizations and businesses*. Generally, the need to make marginal areas visible or/and attractive at a European and national level is at the basis of projects promoting activities for discovery and enhancement of historical, cultural, artistic, and environmental potential through a multidisciplinary approach, with the help of the whole educational community.

Here, the relationship between territory and school implies the birth of a cultural and educational agency, thus contrasting dropout or dispersion in favour of larger school centres.

The interventions for strengthening *basic competencies* offer active experiences that go beyond the traditional and theoretical model of face-to-face learning, but make use of space materialities to review contents, places, and practices. An example is the experience *Our school is a* Maremma Incoming Tour Operator (*M.I.T.O*), which is based on a need to enhance basic skills in linguistic–literary disciplines of children through a teaching method based on *reality task*. For this purpose, the local communities become *hub of contents* and *network of services* able to make available environments, tools, professional figures, and transport services. They contribute to providing information material and documents necessary for carrying out all planned activities, and are involved, together with parents or families operating in tourism or the trade sector, in the creation of a catalogue reporting examples of good practices and teaching scenarios aimed at both enhancing the territorial development and at reducing early school leaving.

The interventions aimed at *inclusion* are consistent with those designed for strengthening basic competence. A *theatre lab* becomes an opportunity to retrain a non-place space and makes it a point of social gathering and cultural offer. Pupils are involved in theatre performance, which helps them enrich their cultural horizon and makes that space an educational materiality useful to experiment relational modalities based on cooperation.

Strengthening of cultural, artistic, and landscape *heritage education* looks at the territory as a common good, potential for *democratic development*, through laboratory teaching practices. Locals can provide specific help, tools, and personal competencies. Families, epitomizing experience and historical memory of their territory contribute to developing a generational interchange of ideas and fostering the integration of foreigners. *Piccolo è bello* (small is beautiful) is a characteristic teaching module focusing on the assumption that municipalities of marginal rural areas may be vital places, given the presence of an active, democratic, multi-ethnic citizenship, and social cohesion.

The aim is to help children reflect about their local community and history as a real laboratory for observation, analysis, interpretation, and reflection while promoting the specific multidisciplinary and interdisciplinary nature of knowledge, by relating the various areas of the curriculum.

With its nine school complexes, **I.C. Satriano di Lucania (PZ)** represents a reference for a geographical area that is rather complex from a cultural perspective, and heterogeneous from a social and economic standpoint. A large number of children live in isolated houses and rarely have occasions to meet with contemporaries. Due to the absence of local structures and associations, they *live most of their extra-school time in the street* and often engage in conflicting behaviour with others. Only a few spend part of their free time in organized facilities. The school intends to intervene for fighting pupils' poor interest, little motivation and passive presence at school that produce failure and are symptoms of school dispersion. The institute has invested in a project of *open and guiding school* in extracurricular time or in summer periods, encouraging exchanges within the local community. It is not only a matter of welcoming children into a more extended time, but of *giving them more opportunities to understand their inclinations and potential.* The environment is part of educational materialities which the school invests in to respond to the idea of *quality school for all, with no exception*: A stimulating and open space, rich in corners / laboratories, enhanced by relationships with local organizations and associations. With regard to *basic competencies*, the action *Life Skills* recalls an educational vision of the school as an environment in which people are welcomed, integrated into, and encouraged to develop their potential at their best, in a climate of well-being and cooperation. New technologies are considered a key tool to promote equality in educational opportunities, especially for linguistic and scientific disciplines.

The projects dealing with social *inclusion* and fight against disadvantage propose a series of modular extracurricular courses on topics connected to pupils' interests and inclinations to be held in welcoming, usable, flexible spaces, with multifunctional areas, equipped with technologies, and designed to foster learning and to enhance talent. The *laboratories* are environments in which actions are internalized and mentally performed, and where pupils may reflect, think, and acquire deeper awareness, learn by doing, and together (peer-to-peer; brainstorming, role playing). An example is *Opificio delle @rti*, an innovative and original *atelier* that enhances pupil's different abilities through a wide programme of activities devoted to *arts* education. The project proposal gains the precious collaboration of public and private partners: Art masters, professionals, experienced artisans, relatives, associations, local government, contributing to qualifying *school as an integrated community space.*

The interventions proposed under the notice related to *heritage education* are aimed at enriching cultural heritage by integrating laboratory spaces inside and outside the school, according to modern outdoor education strategies. Young students become guides who show wall murals, cellars, and sculptures. Here spaces, houses' painted facades, rocky temples' friezes, stony paths of cellars marked by sculptures are "adopted" and offered for tourist routes, fostering historical, linguistic, and orographic competencies.

In the Lucania area, **I.C. 'J. Stella' – Muro Lucano (PZ)** has 13 school complexes situated in many municipalities. The school accommodates children from difficult family backgrounds, poorly motivated and without any learning guidance at home. The school intends to involve local community in a *process of pupils' growth and orientation to professional self* to combine culture and heritage enhancement.

In the area of *basic competencies*, this vision is expressed in the *mini GUIDE in TOUR* project, where pupils are involved in creating touring itineraries that respond not only to specifically disciplinary goals but also to sociocultural and emotional–relational growth of tomorrow's citizens. Students will be able to sort and organize contents, creating dedicated historical–cultural and/or religious paths. Such a mode of operating helps motivate pupils and encourages them to develop their entrepreneurial and creative skills.

Focusing on *inclusion*, *La scuola...il Mondo* is a project designed to strengthen natural attitudes of pupils with learning difficulties, to increase their self-esteem and confidence. Through educational workshops in which school spaces extend to village places (square, cloister, sports fields), each one records his/her growth, measurable not only in terms of content, understood as enrichment of curricular paths, but also at a relational level. Confronting with others, everyone may consciously tell what they know and are able to do.

The project selected in response to the notice aiming at *cultural heritage* supports the creation and adoption of paths for the *protection of cultural goods*. The research work on several sites/artifacts is meant to be carried out in the classroom initially, with the support of multimedia tools, and to be continued later both through educational visits to places of interest and through online environments for production and publication of tourist catalogues and content/itineraries in various languages. The path will result from a synergistic convergence between adults' experience, representing the *past*, young people's abilities, standing for the *future*, and local entrepreneurship (Municipalities, State Police, Civil Protection, Small Museums) as a connector.

I.C.F. P. Polizzano – Gangi (PA) including seven school complexes in an area that in recent years has seen an increase in the number of tourists being attracted by artistic, cultural, and environmental beauties. Nevertheless, the village is marginal both for its geographical position and poor road network. In this context, pupils have few opportunities to broaden their horizons and enrich their cultural background. It is a duty of the school, therefore, to take into account needs of pupils, families, and community. The collaboration with local government public bodies (cultural associations, tourist associations and academies, performance theatres) allows the creation, in the context of *basic competencies*, of recovery paths for Italian and foreign languages based on exploration, dialogue, and teamwork – all elements equally capable of driving actions for social inclusion and against disadvantage. Interventions such as *Racconti dentro la tela* (Stories inside the canvas), on behalf of cultural *inclusion* and fight against disadvantage, promote reading, and strengthen disciplinary competencies. Pupils are taught how to approach the production of artefacts in miscellaneous and unconventional spaces (art gallery, municipal gymnasium, sports facilities, etc.) making use of technologically innovative instruments. This approach recovers the concept of "great narration" (Smith, 2001) and educates the necessity of being together.

Interventions in the field of cultural heritage, in which inclusion is complemented with processes of basic competencies strengthening, are meant to activate a cultural promotion of the Madonie areas as a laboratory of social integration. Outdoor, cooperative, and digital learning characterize these paths. They stimulate the growth of new *Cicero*, guides able to operate concretely in their community, and learn from it and from active institutions – especially confraternities – which are significant for keeping traditions alive, taking care of artistic and cultural heritage, and helping those living in difficult conditions.

I.C. Valle Del Fino (TE) is structured into 15 school complexes distributed over 5 municipalities characterized by different environments, either historical, with rural–agricultural origins, or working-class areas. The area has a low population density, which, in recent years, has constantly decreased, due to poor road network, job unavailability, and seismic events.

With such a background, the school is the main institution to cope with commitment and responsibility of children's education, in cooperation with local institutions and organizations. Municipal government administrations are in charge of paying for utilities during extra-school hours and summer, and will make public places available for events and seminars. Other associations will provide professional operators for educational visits to places of interest, offering both teaching material and expert personnel.

To foster *basic competencies, the project proposal consists of a creative use of multimedia tools (blogs and web pages) to describe territorial backgrounds, through practice and strengthening of mathematics and language ability, and through the creation of concept maps.*

The interventions supporting social *inclusion* and fighting against disadvantage boost the expansion of curricular paths with actions to reinforce basic competencies. Part of the intervention is the creation of a *parenting school*, a *place* that may provide with opportunities for mutual support among adults that share competencies and new educational practices. The involvement of municipal government administrations and associations shares the vision of *school as an educational community*, which takes its roots from the school walls but extends to the whole context it operates in.

Finally, about *cultural heritage*, the school's responses are a proposal of cultural, social, and sustainable tourism and curricular contents supplementary to the school programme. During guided tours, pupils, adopting the *Bring your own database* method, will work to create photographs and photometric captures through free applications available on their own device. Design and writing activities will be performed according to *challenge-based learning*. Exploiting new generation software, the school gives strength to an action built on the invention-based learning, a meeting point between project-based learning and reality tasks, getting pupils involved in the design of solutions to real problems through inventions. From the point of view of technological equipment, the project requires the construction of classrooms designed for making, coding, and 3D printing, with LAN / WLAN connection to access services via mobile devices. The implementation of *talking* models of main local monuments, the installation of a permanent photographic exhibition, and the creation of a website to promote tourism in the area are only a few examples of feasible activities.

I.C Foglianise (BN) includes eight small schools, with the presence of forms of socio-economic and cultural disadvantage. Not only parents but also citizens see the school as one of the few landmarks constantly present, by statute and vocation, in the fortunes of Valle Vitulanese's territory. The school has been acting as a space for community aggregation, extracurricular activities, and meetings aimed at enhancing competencies and sharing educational choices with families.

Concerning *basic competencies*, the proposed project focuses on expressive languages such as the theatrical one (e.g., the module *Theatreland*), in which children may recognize themselves as an active and proactive component of the territory they live in.

The intervention about *inclusion* is purposed to orienting and re-orienting pupils' life project, strengthening operational and active teaching, responsibly involving their family, and opening *positive interactive spaces*. Conventional interior spaces will be transformed into laboratory classrooms, where educational strategies and innovative methodologies will be experimented (cooperative learning, action research, circle-time, brainstorming, learning by doing); unconventional spaces (council hall, assembly hall, etc.) will be re-proposed as new physical and mental spaces. An example of this is the *summer straw atelier*, performed in a laboratory specially set up inside the school for pupils to work with the grain offered by local farmers.

The proposal presented in response to the notice concerning *cultural heritage* deals with the re-appropriation of buildings or monumental works, so as to preserve the memory and testimony of the complex history of these lands, which are fundamental elements of a community's identity.

I.C. di Edolo (BS), structured into 11 complexes across several municipalities, is located in a large and disjointed territory. It records a tendency to social isolation and seems unable to take advantage of diversified cultural opportunities.

The actions for strengthening *basic competencies* mainly focus on language learning, both Italian and English, and on mathematics, through modalities that stimulate pupils' curiosity and improve their engagement using gamification and situational approaches to *act competently*.

Concerning *inclusion*, the projects *Aggiungi un banco in aula. C'è un amico in più* (Add a desk to the classroom for a new friend) and *Insieme ci si aiuta* (United, we help each other) are paths aimed at building collaboration, mutual support, and acceptance. Municipalities and local government public bodies work together for a close and real correlation between school and territory, in which children are recognized as active and proactive members of the community they live in.

This collaboration is shown more consistently in projects covering the *cultural heritage* notice, where places vested with historical memory become educational ways and objects for constructing historical and disciplinary knowledge. Country churches and natural parks extend classes and become the subject of an in-depth study and the development of digital curriculum contents related to cultural heritage (Open Educational Resources).

Conclusion

As often suggested by Roehl (2012), it is important to identify strategies that give us the opportunity to analyse how materials participate in the educational action by shaping and being shaped by the practices themselves.

In section 'The cultural framework of educational materiality in small schools' of this chapter, we have identified some dimensions of materiality that let us consider objects as *actors*, thus trying to make their contribution to the practice more visible, and putting the accent on what "they allow to make or not to make happen" (Sørensen, 2009). Such an identification helps us to better "disarticulate" the action spaces (Roehl, 2012) by making the reader aware of what happens in the classroom (intended as an educational environment extended to all places included within the didactic planning) is the result of ways of intertwining and combining objects and sites (McGregor, 2004).

Interventions in support of basic competencies, social inclusion, and combat against hardship find in materiality a key to interpreting the "real *datità*" (Massa, 1997) through a dialectics possible among (often latent) devices that are part of the field of educational experience (Barone, 1997).

In the conclusion, we pull the strings of a proximal reading of the materiality present in the projects of selected small schools, also bringing attention to the notice on innovative environments and trying to identify in digital technology a lever of change and innovation incorporated in the pedagogical practices.

The materiality that emerges from the educational vision expressed by the I.C. Tozzi is the one that best recalls *forms of educating cities for widespread and inclusive learning* by strengthening of centrifugal educational formulas in case of difficulty. Interaction with territory (home, public places, recoverable spaces) is the educational relationship that runs through all school's planning and is enhanced by technology. The *task of reality (authentic task),* therefore, uses a classroom enhanced through Cloud computing, to provide children with home access to the classroom whereas they are not able to reach their school because of impeding weather conditions or lack of public transport. This system supports projections that can be viewed on the blackboard or on single devices as a production tool available to be used by groups, even remotely. The classroom is organized in mobile laboratories and disciplinary guides to guarantee active learning (based on problem-solving and hands-on activities) and a dynamic relationship with the territory as a hub of content, space, and services.

The I.C. Satriano di Lucania proposes a vision of a school as an integrated centre connecting spaces internal and adjacent to the school. In this context, technology enables forms of study engagement, thus recalling the *Mechagogical dimension* of materiality. The use of TabletBook with document camera allows teachers and pupils to share whatever object built or retrieved on the screen of interactive whiteboards (IWB), as well as directly create videos for a rewarding and adaptive learning.

The I.C. J. Stella promotes pupil's growth and guidance to professional self through *a materiality resulted from the relationship between school and territory in which the city provides educational opportunities*, and workshops where local community spaces are combined with a space for mobile learning available in the classroom and aimed at learning by doing.

The I.C.F. P. Polizzano reviews material resources to support educational courses that combine outdoor training, cooperative learning, and digital learning. *Material forms of the school and of adjacent spaces are re-mediated to support basic skills learning*, also giving new value to their territorial context for creating a *community being partner in educational choices*. This approach, oriented to the production of artefacts in a mixed articulation of spaces, extends the classroom environment and makes it virtual through a Virtual Learning Environment (VLE) and Learning Management System (LMS), supporting remedial education and differentiated learning.

In the I.C. Valle del Fino, classrooms are enhanced by networking and Cloud computing, with a material *dimension of Meccagogia* to allow access to didactics, and, above all, to remedial education even for children experiencing homebound situations, logistic, and transport difficulties often found in small schools. Furthermore, with content sharing and videoconferencing systems, it is possible to support pupils' learning, regardless of their physical presence in the classroom, thanks to the possibility to record and store lessons. This allows these pupils to feel included in the teaching–learning process and in their class dynamics, thus promoting small school as a digital school centre.

The projects presented by the I.C Foglianise (BN) recall two axes of materiality: One *re-mediated, to support curriculum by competencies,* and the other *assisted by forms of Meccagogia,* as shown by the proposals on innovative environments. Laboratory activities are supported in multimedia learning environments, functional to the implementation and achievement of significant results, namely, skills specifically enabling exploration, phenomena classification, definition of issues and problems, setting and understanding of connections, construction of new interpretative scenarios, and design of solutions. Technology, therefore, becomes material element that helps review space and traditional objects (contents, activities) making them connected and interindividual, and strengthens them in terms of exploration applied to scientific disciplines and *makering.*

The I.C. Edolo with the primary objective of improving pupil's educational success, promotes a *collaborative classroom, capable of supporting forms of peer-to-peer learning* also vertically, finding a valid support in the *material dimension of Meccagogia*. Interactive whiteboard (IWB) technology is enhanced with DViT, creating an interactive multi-touch, multi-gesture, quad-touch (supporting up to four contemporary touches on the work area) interactive work surface working with finger input, *pen-like* or any pointing device. The path that each institute is undertaking to translate its project has its central point in an idea of school intended as an *ecosystem of material devices* for teaching. All the investigated experiences emphasize innovation through the use or reuse of spaces that break the traditional order of a class through disciplinary knowledge and a revisited school time in which human and material elements interact continuously.

To verify the implementation of the school projects described above, it will be necessary "to follow the protagonists, those who, influencing it, really reconstruct it", as Sorensen argues, so as to critically evaluate how materiality is integrated into teaching practices.

Notes

1 Sections "The cultural framework of educational materiality in small schools", "Qualitative analysis of projects", "Conclusion" were written by Giuseppina Rita Jose Mangione. Sections "The opportunities of 2014–2020 National Operational Programme", "Small schools profile for an educational materiality: a quantitative analysis" were written by Samuele Calzone. In the framework of the European Structural Funds 2014–2020, and in the specific context of the National Operational Programme "Per la Scuola – Competenze e Ambienti per l'apprendimento", this contribution is the result of a scientific cooperation involving two actions: (i) "Monitoraggio, ricerca e supporto didattico (GPU 2014–2020)" – project code 4.1.4 A-FSEPON-INDIRE-2015-2, and (ii) "Piccole Scuole", project code 10.1.8. A1-FSEPONINDIRE-2017-1.

2 http://piccolescuole.indire.it/il-progetto/

3 http://piccolescuole.indire.it/il-progetto/manifesto/

4 The Small School project, financed under the 2014–2020 multi-fund National Operational Programme "PON for schools: competencies and learning environments" (2014IT05M2OP001) under the umbrella of Axis I "Investing in Education and Lifelong Learning" promoted the creation of a national network or movement, counting at the date over 200 schools based in geographically isolated areas. Movement's website at: http://piccolescuole.indire.it/

5 As for the educational interventions (ESF Fund), the notices examined are the following: Notice n. 10862/2016, Notice n. 1953/2017, and Notice n. 4427/2017.

6 Programme 'PON Per la Scuola 2014-2020', p. 2, www.istruzione.it/pon/ilpon.html

7 In Position Paper, p. 6: 'In terms of 2014–2020 programming, five Southern Regions are classified as "less developed" (GDP per capita <75% of the EU average), three as "intermediate" (GDP per capita between 75% and 90% of the EU average), and no region as "more developed" (GDP> 90% of the EU average)'.

8 The civic centre is conceived as a space to deliver educational as well as recreational, sporting, and social activities, which aim to raise awareness on issues of legality, sustainability, climate change, etc.

9 The governance system 'Joint Programming Management (Gestione Unitaria del Programma (GPU) 2014–2020)', is the online system for the management, monitoring, and documentation of the activities of the 2014–2020 National Operational Programme – 'For School', http://pon20142020.indire.it
10 Management Authority (Italian: Autorità di Gestione – AdG) is an authority of the Ministry of Education, University and Research (MIUR).
11 Notice n. 12810, p. 8.
12 Notice n. 12810, p. 3.
13 Ibidem.
14 Notice n.10862/2016, p. 2.
15 Pupils were allowed to participate in more training sessions; specifically, many of them already involved in projects referred to Notice n.10862/2016, took part in initiatives offered by projects under Notice n.1953/2017.
16 "More Developed Areas": Emilia Romagna, Friuli Venezia Giulia, Lazio, Liguria, Lombardia, Marche, Piemonte, Toscana, Umbria, Veneto. The Programme considers also regions "in Transition Areas": Abruzzo, Molise, Sardegna.
17 The term "inland areas" refers to territories characterized by: a significant distance from main urban centres offering essential services (health, education, collective mobility); high availability of vital environmental resources (water resources, agricultural systems, forests, natural and human landscapes) and cultural resources (archaeological heritage, historical settlements, abbeys, small museums, art and craft centres); complex territories, due to natural systems, dynamics and human settlement, and depopulation processes that have characterized them.
18 It is a national network of Small Schools whose purpose is to enhance their educational and cultural function to combat depopulation, http://piccolescuole.indire.it/il-progetto/manifesto/
19 Higher grades do not strictly depend on pupils' participation in the Programme; participating in the PON, together with other elements not covered in this work, helps generate a climate of scholastic well-being that fosters better results.
20 The Manifesto can be downloaded at: http://piccolescuole.indire.it/il-progetto/manifesto/
21 Project: "Small Schools" – carried out under the 2014–2020 multi-fund National Operational Programme "PON for schools: competencies and learning environments" (2014IT05M2OP001) under the umbrella of Axis I "Education" – OS/RA 10.1 "Reduction of early educational failure and of scholastic dispersion." – Azione 10.1.8 / Progetto: "Piccole Scuole" – Codice progetto – 10.1.8. A1-FSEPON-INDIRE-2017-1 – CUP – B59B17000010006.

References

Baldascino, R. (2018). Le inquietudini della scuola del futuro (Concerns of the school of the future). *Rivista dell'istruzione, 5*, 79–82.

Barone, P. (1997). *La materialità educativa (Educational materiality).* Milan: Unicopli.

Berghman, J., & Salais, R. (2001). *Social exclusion and European policy.* Cheltenham: Edward Elgar. 27–46.

Bouck, E. C. (2004). How size and setting impact education in rural schools. *Rural Educator, 25*(3), 38–42.

Campagnoli, G. (2018). Il disegno di una città educante (The educating city project). *Rivista dell'istruzione, 5*, 74–78.

Dewey, J. (1970). *Scuola e società (School and society).* Florence: La Nuova Italia.

Edwards, R., & Carmichael, P. (2012). Secret codes: The hidden curriculum of semantic web technologies. *Discourse: Studies in the Cultural Politics of Education, 33*(4), 575–590.

Fenwick, T., Edwards, R., & Sawchuk, P. (2015). *Emerging approaches to educational research: Tracing the socio-material.* London: Routledge.

Fenwick, T., & Landri, P. (2012). Materialities, textures and pedagogies: Socio-material assemblages in education. *Pedagogy, Culture & Society, 20*(1), 1–7.

Fenwick, T., Nerland, M., & Jensen, K. (2012). Sociomaterial appróaches to conceptualising professional l learning and practice. *Journal of Education and Work, 25*(1), 1–13.

Ferrante, A. (2016). *Materialità e azione educativa (Materiality and educational action).* Milano: Franco Angeli.

Gorur, R. (2011). Policy as assemblage. *European Educational Research Journal, 10*(4), 611–622.

Harrison, D. A., & Busher, H. (1995). Small schools, big ideas: Primary education in rural areas. *British Journal of Educational Studies, 43*(4), 384–397.

Johnson, R. B., & Onwuegbuzie, A. J. (2004). Mixed methods research: A research paradigm whose time has come. *Educational Researcher, 33*(7), 14–26.

Kvalsund, R., & Hargreaves, L. (2009). Reviews of research in rural schools and their communities: Analytical perspectives and a new agenda. *International Journal of Educational Research, 48*(2), 140–149.

Landriscina, R., & Viteritti, A. (2016). Sociomateriality in the classroom: Practices of didactic innovation. *Scuola democratica,* 7(1), 93–116.

Latour, B. (2006). Where are the missing masses? The sociology of a few mundane artefacts. In, W. E. Bijker & J. Law (Eds.), *Shaping technology/building society: Studies in sociotechnical change* (pp. 225–258) Cambridge, MA: the MIT Press.

Lawn, M., & Grosvenor, I. (Eds.). (2005, May). *Materialities of schooling: Design, technology, objects, routines.* Oxford: Symposium Books Ltd.

Mangione, G. R. J., & Calzone, S. (2019). The Italian Small School toward smart pedagogy. A cross-reading of opportunities provided by the National Operational Program (PON) "For schools 2014–2020–skills and learning environments". In *Didactics of smart pedagogy* (pp. 233–252). Cham: Springer.

Mangione, G. R., Calzone, S., & Bagattini, D. (2017). Digital environments for Small Schools. The enhancement of laboratory spaces within a renewed classroom concept. *Form@ re-Open Journal per la formazione in rete, 17*(3), 83–100.

Mantegazza, R. (2001). *Unica rosa: cinque saggi sul materialismo pedagogico (Unique rose: Five essays on educational materialism)* (Vol. 2). Milano: Mimesis.

Marcarini, M. (2018). "Fare scuola" in relazione attraverso gli "spazi colore" realizzati al civico polo scolastico "a. manzoni" a milano: un'alternativa all'aula scolastica per proporre nuove metodologie didattiche. ("Making school" through relationships, using "colour spaces" created at the civic school "A. Manzoni" in Milan: An alternative way to traditional classroom proposing new teaching methods). *Quaderni GRIMeD,* 42.

Martin, S. H. (2002). The classroom environment and its effects on the practice of teachers. *Journal of Environmental Psychology, 22*(1–2), 139–156.

Massa, R. (1997). *Cambiare la scuola: educare o istruire? (Changing the school: Educating or instructing?)* Roma: Laterza.

Masschelein, J., & Simons, M. (2013). *In defence of the school. A public issue.* E-education. Leuven: Culture & Society Publisher.

McGregor, J. (2004). Spatiality and the place of the material in schools. *Pedagogy, culture and society*, *12*(3), 347–372.
Mitchell, D. (2008). *What really works in special and inclusive education. Using evidence-based teaching strategies*. New York: Routledge.
Mulcahy, D. (2011). Teacher professional becoming: A practice-based, actor-network theory perspective. In *"Becoming" a Professional* (pp. 219–244). Dordrecht: Springer.
Nespor, J. (2011). Devices and educational change. *Educational Philosophy and Theory*, *43*(sup1), 15–37.
Paechter, C. (2004, March). Power relations and staffroom spaces. *Forum for Promoting 3–19 Comprehensive Education*, *46*(1), 33–35.
Patriarca, L. (2018). Evviva l'intervallo (e il cortile) (Long live the interval (and the Court yard). *Rivista dell'istruzione*, *5*, 43–45.
Pontecorvo, C. (2000). *Manuale di psicologia dell'educazione (Handbook of educational psychology)*. Bologna (Italy): Il Mulino, p. 420.
Postman, N. (1979). Teaching as a conserving activity. *Instructor*, *89*(4), 38–42.
Roehl, T. (2012). Disassembling the classroom–an ethnographic approach to the materiality of education. *Ethnography and Education*, 7(1), 109–126.
Semeraro, R. (2014). L'analisi qualitativa dei dati di ricerca in educazione (The qualitative analysis of research data in education). *Italian Journal of Educational Research*, *IV*(7), 97–106.
Smith, J. K. (2001). A little story about Metanarratives: Lyotard, religion, and postmodernism revisted. *Faith and Philosophy*, *18*(3), 353–368.
Sørensen, E. (2009). *The materiality of learning: Technology and knowledge in educational practice*. Cambridge: Cambridge University Press.
Strike, K. (2008). Small schools: Size or community? *American Journal of Education*, *114*(3), 169–190.
Thaler, R., & Sunstein, C. R. (2014). *Nudge. La spinta gentile: La nuova strategia per migliorare le nostre decisioni su denaro, salute, felicità (Nudge. Improving decisions about health, wealth, and happiness)*. Milano: Feltrinelli Editore.
Turrisi, M. R. (2018). Le pareti di una scuola raccontano (School walls speak). *Rivista dell'istruzione*, *5*, 33–35.
Wentzel K. R. (1997). *Student motivation in middle school: The role of perceived pedagogical caring. Journal of Educational Psychology*, *89*(3), 411–419.

Documents concerning National Operational Programme (PON)

Notice no. 12810/2015, http://www.istruzione.it/pon/avviso_ambienti-digitali.html
Notice no.10862/2016, http://www.istruzione.it/pon/avviso_inclusione.html
Notice no.1953/2017, http://www.istruzione.it/pon/avviso_competenze-base.html
Position Paper, http://www.istruzione.it/pon/ilpon.html#sec_pro
Programma PON Per la Scuola 2014–2020 (National Operational Programme "For School"), http://www.istruzione.it/pon/ilpon.html

Chapter 8

A proposal of smart pedagogy for the school

Educational curation as an active methodology

Marilene Santana dos Santos Garcia and Luciano Frontino de Medeiros

Introduction

This chapter intends to expand on the definition of educational curation, and what we can understand from the theoretical–practical perspective of active methodologies. The content within this study is related to empirical, theoretical, and reflexive research methods that were carried out in classrooms of students from the course 'active learning and hybrid learning', a class taught in master's degree programmes in Brazil.

To begin, there are some correlations that should be made clear: (i) Working with active methodologies of which innovate practices are one of the actions of smart pedagogy; (ii) educational curation promotes a relationship between teaching and learning with the current events of the world, and searches for more engaged, critical, and active students; and (iii) the curator teacher is one who is continually improving himself, seeking new resources, inside and outside the school, and values research proposals appropriate to his students.

The present chapter, as a result of the empirical research of methodological experiments, shows some orientations about its abilities, competences, and digital fluencies. The authors used as references in this study are Bacich and Moran (2018), Garcia (2017), Garcia and Czeszak (2019), and Vickery (2013), among others.

What is educational curation?

The term curation, or curatorship, is related to the cultural universe, wrapping events with works of art, and is related to the curator of exhibitions. This curator is the one who selects and organizes works of art for a particular event or exhibition. The concept of an educational curator can already be observed in this conception, which has the need to search for a focus that wants to pass a message to be captured by its students. This message covers curating, caring for, organizing, orienting, and questioning certain senses. As a result, the concept of the arts can be used to approach the educational curator in the practices that are carried out in the school (Garcia & Czeszak, 2019).

According to Černý (2017, p. 1), "Etymologically, the concept of curatorship is based on care or administration, the curator is a guardian, someone who supervises the content or property entrusted to it and works with it to bring benefits and to be safe".

With the increasing presence of digital technologies in the field of education and the high volume of information available from different digital platforms, educational curation has become increasingly relevant and necessary (Černý, 2017). Because of this, curators should focus on the selection and organization of content related to the most diverse areas of knowledge while also warning students about fake news and unreliable sources. What is important is that the curatorial project should be a negotiable instrument for the students in search of new knowledge.

The educational process goes through methodologies that broaden students' discernment. These include how to construct arguments, how to make good source choices, and how to use and expand content in school tasks. Within this context, one can recognize the functions of teachers engaged in co-authorship and the process of filtering, refining, and reusing information.

Whether conscious of this role or not, the teacher has always exercised, to a greater or lesser extent, the role of a curator. However, the profile of an educational curator emerges more emphatically in a scenario of needs imposed by new demands for their practices, which imply domains and digital fluencies, so that they can dialogue pedagogically, cognitively, and affectively with current students.

The role of the educational curator is not only to face the amount of information in their practices pedagogically, but to elaborate plans and activities that can allow bases for knowledge to be expanded, co-produced, reworked, and reflected upon critically.

Digital fluencies of the educational curator

According to Mihailidis and Cohen (2013) and Černý (2017), the teacher should excel in the digital fluencies and skills that are shown in Tables 8.1–8.3.

Digital fluency bridges the gap between technologies that exist in the world and that are accessible. They are then transferred to the context of teaching and learning, creating value for them. The prevailing idea is that the teacher can consider their individual experiences outside of school with new media and cutting-edge technologies, promoting new meanings for their professional scope.

Digital fluencies should be considered as mandatory knowledge for the teacher who wants to work with a smart pedagogy, whether it involves digital curation or not. In turn, the teacher needs to reflect and practice transpositions of digital fluency that already dominate in the social sphere, such as using social networks and applications in their daily lives so that they can strengthen their professional skills in school environments. If there is low digital fluency, the teacher should also seek help from other colleagues and their own students, exchanging experiences on such aspects.

Table 8.1 Digital fluency for educational curation

Digital fluencies	*Description*
Active project methodology	Using available technologies for both the development and presentation of project results.
Be a continuous digital and educational apprentice	Finding references to new applications and technological innovations available for the educational profile of the students and school.
Office package domain	Using the tools of the Office package, mainly the use of text, spreadsheet, and presentation editors.
Affinity with technology inside and outside the school environment	Exercising the use of technology and demonstrating this ability in several activities.
Know game and gamification differentials	Exploring games and game activities to enrich classes, promoting lucidity.
Incorporate self-taught skills	To understand that self-learning is a process, the improvement of educational work must start from the teacher to positively reach the students.
Seek innovation in mobile methodologies	Delving into the possibilities of mobile device practices and integrating them into blended learning activities.
Work perception of reality and seek solutions to problems	Researching current digital thematic contexts to broaden student referrals in project proposals.
Develop collaborative skills	Understanding collaboration in various senses, such as sharing information, writing texts in co-authorship, discussing divergent ideas, etc.
Use social networks in everyday life	Having a personal experience with social networks and conditions to critically extend this knowledge to the educational scope.
Increase knowledge of current technology to make use of them in educational plans of curatorship	Talking to colleagues and other experts to find out what is new and selecting the best technology for the educational field.
Know how to use cloud storage and other related devices	Understanding how cloud storage (for example, Google Drive) works and how to use the tools for it. This optimizes hardware devices.
Explore usability concepts	Exploring ways to explain the use of technological tools to perform some specific task. This is helped by proposing some curated research to your students.
Get to know sites/blogs/applications that analyse fake news	These are websites, blogs, and digital curatorial and journalistic applications that provide guidelines on how to analyse fake news.

Source: The authors.

Skills of the educational curator

Uskov et al. (2018) state that a number of teaching and learning strategies have emerged in recent decades, but we must also invest in research on "those innovative approaches supporting the main concepts of Smart Education, Smart Classroom, and Smart Pedagogy", that is, their "smartness". Thus, they seek important elements of this practice's effectiveness in requirements such as: (a) Adaptivity, (b) sensing, (c) inferring, (d) anticipation, (e) self-learning, and (f) self-organization.

Lorenzo and Gallon (2019) understand that the current moment is aimed at a new pedagogy, but this should not mean offering technologies that present the knowledge of society, but instead seek to engage the students in their learning with different practices in search of knowledge. The authors emphasize that these practices should be more reflexive, dialogic, and engaged, in which different competencies can be worked out, such as instrumental competence, interpersonal skills, and systemic competencies.

From these observations, one can infer that the educational curator must also have a range of skills, which need not be immediate or complete, but must be addressed as their demands for pedagogical work take place. Thus, it is understood that smart pedagogy should function under this framework, in which attention should be paid to the new teaching and learning processes and the teacher should be prepared to face them.

Table 8.2 presents a number of skills needed by the educational curator. It is a reference framework, so it should not be seen as including exhaustive information on the subject.

Allied to the digital fluencies and skills described in Tables 8.1 and 8.2, the teacher, as the curator, will be able to demonstrate some required skills for this function.

Table 8.2 Skills required for the educational curator[a]

Skills required	*Description*
Customize educational methodologies	The educational curator has the ability to customize teaching in the sense of proposing pedagogical plans and specifically curated projects to their students.
Expand the use of projects with gamification methodologies	The concept of gamification should be incorporated by the curator to insert playful elements inspired by the games, whose language is known by the students.
Guide the creation of research opportunities in the context of school learning	The educational curator has the ability to create opportunities for students to investigate the theme.

Skills required	*Description*
Knowing the configurations of the students' mobile devices	It is necessary to be up to date with the configurations of the students' mobile devices to be able to propose appropriately curated projects with their use.
Analysing the context and environment of the target audience	Understanding the reality of the school's location prompts the emergence of environmental themes with proposals of student interest, etc.
Critically analyse the use of social networks	Social networks can be grounds for educational curatorial projects that can be analysed critically by teachers and students.
Innovate forms of interaction with the students by using feedback	Practising more effective communication with students. Feedback should be effective and constructive.
Teach project structure	Learning the systematics of projects – start from a problem situation, propose a solution, and apply the results.
Systematic research work	Do not conduct an investigation beyond the condition and ability of the students; in this sense, students should curate project research advances from their initial level.
Having the ability to manage different points of view	It is an educational act to curate different opinions on certain subjects and to teach students how to derive the best results from a diversity of positions.
Practicing versatility in lesson proposals	Curation expands students' research references so as to give versatility and variety to research proposals.
Having the capacity for interpretation and analysis	The student on an educational curriculum must develop the capacity for interpretation and textual analysis to develop their own points of view.
Practicing studies on educational design	The curator should be interested in the presentation formats of both their content and that requested of the student.
Working with professional enhancements	Becoming a curator of content, resources, and referrals makes the teacher improve professionally.
Knowing the project structure	Mastery over the structuring of projects makes the curator have the security to heal the elements of each of their phases.
Incorporate the role of researcher	The curator incorporates the job of researcher but also knows the limits necessary to work with their students in terms of topics of interest and systematic searches.

(Continued)

Skills required	*Description*
Incorporate convergent and divergent forms of thinking	Elaborating curated projects that work on divergent and convergent thinking to support the diversity of ideas.
Be creative	Extending creative processes targeting both students and the curator themselves.
Questioning the practice itself	The idea of curation questions practices rooted in the content domain that should be reviewed, giving them broader senses regarding learning.
Transpose ideas into other contexts in which there are little financial resources, etc.	It is important to work on ideas that can be reworked in other contexts. For example, a mathematical calculation can give a better quality of life for those with little financial resources.
Openness to new challenges	The curatorial project is challenging and demands advances not only in terms of methodologies, but also in principles about what it is to teach in an engaging and participatory manner.

Source: The authors.

a These skills required for the educational curator are part of the research of the master's degree in education and new technologies programme at UNINTER.

General competencies for the educational curator

Table 8.3 shows the required competencies for the educational curator.

The educational curator should practice having more engaging dialogues with students and other fellow teachers while working within the school context. The innovative role of this teacher includes making timely educational solutions with well-chosen resources that meet the profile needs of the students, who are actively interested in the learning (Vickery, 2013).

As a result, there is a need to have educational training for teachers. This is mainly because students should be guided to carry out research with the best sources; to bring information punctuated by their interests; to demonstrate critical sense without their productions; and, finally, to compare and differentiate real news from fallacious stories (Bacich & Moran, 2018). The student should also reflect on how communication is conveyed and prepared, including nuances and intentions behind chosen expressions (Prisloo, 2019).

This curator teacher thus becomes a continuous learner, observing the transformations of the present world, and elaborating them as challenges to

Table 8.3 Required competencies for the educational curator[a]

a	Plan classes that contemplate the methodology of projects, clarifying their purposes and tools.
b	Have knowledge of digital technologies to be able to converse with students about the construction of the curatorial project.
c	Be the author of texts, presentations, links, blogs, etc.
d	Offer stimulus-rich environments, such as using suggestions from research sources.
e	Incorporate knowledge about active learning and learning styles of students within the classes.
f	Suggest gamification activities as specific tools for this purpose.
g	Transfer forms of autonomy and self-learning practised in curated projects to the students.
h	Prepare specific activities for students' mobile devices by knowing the limitations of the configurations of these devices.
i	Work in plenary sessions to present projects directly related to the students' reality.
j	Present tools for co-authorship that favour collaboration.
k	Practice projects with students that involve the use of social networks to communicate research results, approaching the realities of the students.
l	Look for work differentials that involve interaction processes.
m	Develop the students' projects.
n	Provide access to different references/sources to support the project.
o	Propose investigative actions with and for students.
p	Articulate different points of view and contributions for the treatment of a problem.
q	Make new combinations of what has already dominated in terms of active teaching methodologies.
r	After each experience, check the students' beliefs to continue improving.
s	Transpose and combine different languages for educational design.
t	To not fear being the centre of the contents, but seek recognition in the new place that occupies.

Source: Based on and expanded from Garcia and Czeszak (2019, p. 3).

a These skills required for the educational curator are part of the research of the master's degree in education and new technologies programme at UNINTER.

be proposed to his students. Garcia and Czeszak (2019, p. 24) describe some of these functions, which bring new implications to educational practices.

Practices of the teacher curator:

- Draw lesson plans in which interdisciplinary projects demand that students access digital sources and allow for transformations and creative combinations on their part.

- Develop activities in which the student can act in an autonomous, responsible manner, but at the same time seek interactions with their peers and communicate with them about the solutions found.
- Create a willingness and interest in students to research real problems and thus become responsible for applicable solutions.
- Work on propositions of research paths, in which the purpose is to enrich the supply of researched sources with quality differentials.
- Work on evaluation forms in which the value of the student–researcher is relevant both with respect to the process and also to the determined objectives in their propositions.
- Create visual cartographies, in the form of infographics, on where to find and dispose of information on certain research areas to communicate results to all.
- Involve different school levels in projects where there are necessarily varieties of access and forms of research, bringing new perspectives from proposed challenges.
- Create critical communities about the solutions brought to real problems.
- Teach and learn about how to develop projects and research to develop solutions.
- Make an inventory of the available digital technology tools for support.

From this inventory, we can see that one of the main purposes of the educational curator is to "take care" of or to expand visions about content, not only with respect to the quality of sources, but also with respect to formats, supports, and combinations.

With regard to formats, it is possible to emphasize that the fact of working with more innovative educational models requires direct reflections on what the educational curators know about the students, their maturity to work in a pedagogical dialogue, and their expansion on visions and processes. An important example is blended learning, which presupposes that the educational curator has knowledge about what to propose distinctly and interactively in distance and face-to-face environments, on technological supports, as well as on cognitive abilities, motivations, learning styles, and personalized ways of learning and teaching.

Blended learning, based on Horn and Starker's (2014) concepts, has raised increasingly relevant discussions and reflections on how concrete schools can promote disruptive innovation by effectively absorbing digital technologies at the same time as it empowers students' actions with teachers, defining objective goals that incorporate the differential of digital mediation as a mechanism for access to dispersed content in the world. It is becoming increasingly necessary to develop a focus on quality, given that informational quantity is not lacking.

The digital universe offers a series of tools that enable actions that support some form of educational curation. They can be used for co-authorship, the presentation of content, communication between groups, research, and repositories, among other functions. Some general functions of these tools, which in themselves aim to promote active learning:

General Functions of Digital Tools for Educational Curation

- Creation of mental maps, quizzes, flashcards, etc.
- Creation of animations.
- Creation of stories with the inclusion of various audiovisual resources.
- Creation of infographics.
- Transformation of content into books.
- Development of Webquest/treasure hunt projects – hybrid methodologies.
- Reference organizations.
- Organization of annotations and content searched for by students.
- Development of online assessment processes.
- Organization of annotations and content researched by students and teachers that can be shared.
- Elaboration of narratives with content prepared by the students, bringing a new perspective to reflections on problems and their solutions.

It is thus believed that the greater the extent to which the educational healer can mobilize more affective and effective interests, reasonings, and ties of the students, the more dedication there will be by the students to go one step further.

Learning, in this case, becomes a process of signification and re-signification from new perspectives and views about what is known.

This is one of the tasks of the teacher–curator, who must have commitments to processes of engagement to solve problems and competencies for research.

The educational curator is an organizer of content focused on research and research that uses the choices and involvement of students. He uses logical reasoning, and, at the same time, is also sensitive and intuitive. He knows how to manage digital technological resources to change the realities of everyday school life, with attitudes on and responsibility for the choices of the digital technological resources that will be used.

The educational curator's actions cannot be carried out solitarily, but in solidarity, be it with their colleagues or with their students, who should benefit directly from collaborative and cooperative procedures, which are aspects of great importance for curricular education. Working collaboratively and providing the organized material so that others can contribute to it, thereby becoming richer and more updated, are principles that should be emphasized.

References

Bacich, L., & Moran, J. (2018). *Metodologias Ativas para uma educação inovativa: Abordagem teórica e Prática (Active methodologies for an innovative education: Theoretical and practical approach)*. Porto Alegre: Penso.

Černý, M. (2017). Curating content as a way to promote interesting resources. In *INFORUM, 2017: 23rd Annual Conference on Professional Information Resources*, Prague, May 30-31.

Garcia, M. S. S. (2017). *Mobilidade tecnológica e planejamento didático (Technological mobility and didactic planning).* São Paulo: SENAC.

Garcia, M. S. S., & Czeszak, W. (2019). Caminhos para a prática de curadoria educacional (Paths to the practice of educational curation). *Pátio – Elementary School, XXII*(88), 22–25.

Horn, M. B., & Starker, H. (2014). *Blended learning: Using disruptive innovation to enhance education* (1st ed.). eBook Kindle, San Francisco, CA: Jossey Bass.

Lorenzo, N., & Gallon, R. (2019). Smart pedagogy for smart learning. In L. Daniela (Ed.), *Didactics of smart pedagogy: Smart pedagogy for technology enhanced learning* (pp. 41–69). Cham: Springer International.

Mihailidis, P., & Cohen, J. N. (2013, Spring). Exploring curation as a core competency in digital and media literacy education. *Journal of Interactive Media in Education.* Retrieved March 2019 from https://files.eric.ed.gov/fulltext/EJ1007224.pdf.

Prisloo, P. (2019). (Un)dreaming the future: (Dis)connecting (some of) the dots in digital technologogy and education futures. In *UNISA – EDU4FUTURE Conference*, University of Leeds, 2–3 July.

Uskov, V. L, Bakken, J. P., & Uskova, M. (2018). Smart pedagogy innovative teaching and leaning strategies in engineering education. In *IEEE World Engineering Education Conference*, EDUNINE, Buenos Aires, Argentina.

Vickery, A. (2013). *Developing active learning in the primary classroom* – eBook (1st ed.) https://uk.sagepub.com/en-gb/eur/developing-active-learning-in-the-primary-classroom/book239258

Chapter 9

Future challenges for ensuring technology-enhanced learning

Linda Daniela

Technology-enhanced learning (TEL) has become, to some extent, a mantra that is invoked whenever technology is involved in the learning process, whether it is the acquisition of computer skills or the use of learning management systems or smart devices in the learning process. Educational researchers have focused on TEL from different dimensions, such as (i) analysing the acquisition of specific skills in the use of technology; (ii) analysing ways to promote learning through technology; (iii) analysing how the use of technology affects learning outcomes; (iv) analysing the place and role of innovative technological solutions in the learning process; (v) analysing the pedagogical strategies to be used in TEL; and (vi) analysing future visions from a TEL perspective.

This book has been devoted to ideas for planning and organizing TEL in a school setting. At the beginning of the book, I offered a conceptual framework for pedagogical work in TEL, defining the dimensions that need to be addressed and considered when organizing TEL from the perspectives of related challenges and of human development and teaching.

Divya Menon, Sowmya BP, Margarida Romero, and Thierry Viéville explored the possibilities of computational thinking (CT) development with unplugged activities, learning programming languages, and educational robotics. They defined the benefits and challenges of each type of CT acquisition they analysed, as well as the investment and the prior knowledge required, and came to the conclusion that current practices in the way programming introduced in K-12 education have led to a huge diversity in terms of pedagogical strategies. The authors believe that further research and pedagogical experiments are needed to understand how to better promote CT development through a variety of activities and have initiated several projects in this area, the results of which will be provided in future publications.

Miroslava Černochová, Hasan Selcuk, and Milan Svoboda studied how to develop algorithmic thinking using programming with Scratch. This is a quasi-experimental study wherein 16 specific tasks were developed to advance the development of algorithmic thinking, and the results were tested by pre- and post-activity algorithmic thinking. The authors believe that there is a need for programming at school to develop logical thinking and the ability to design algorithms and programming is important for their lives.

Emmanuel Fokides, on the other hand, presented the results of his research on the pedagogical aspects of using digital games (Microsoft's Kodu Game Lab) in education, and presented results from 11 different projects implemented at the primary school level. All projects used a quasi-experimental research design. The results were obtained through both knowledge assessment tests and student surveys, and Fokides concluded that the use of games is motivating and students gain a better understanding of programming concepts. It was also found that the use of games facilitates active and self-directed learning. The author believes that there is a need for closer collaboration between teachers and an ICT expert to develop various technological solutions since teachers have the pedagogical knowledge and ICT practitioners have the technological knowledge, and such cooperation can lead to the development of higher-quality learning materials.

Next, Polyxeni Kaimara, Sofia-Maria Poulimenou, and Ioannis Deliyannis analysed how 21st-century super skills – critical thinking, communication, collaboration, and creativity – can be mastered at the same time as learning about cultural heritage. Readers were offered the idea of a single project aimed both at supporting the documentation and protection of cultural heritage and, at the same time, promoting the learning process for both communication and educational purposes and to enhance the creation of effective outcomes, like empathy, which are the fundamental aspects of the learning experience. The authors believe that the use of technological solutions and the involvement of individuals in the identification of cultural heritage support the cultural heritage-oriented content and engage learners by giving them control of the process through interaction with the content and other content creators. They conclude that innovative pedagogical work is essential to use the latest technological solutions in an attractive way to promote learning.

A vision of playing in preschool while also using digital games was summarized by Jennifer Wong-Powell. She suggested the possibility to explore the opportunities afforded through digital play at home and in a school environment, and believes that the digital environment offers the opportunity to expand the playground, provides more imaginative play, extends the range of gestures used, and enhances children's experiences. It is possible to diversify modes of interaction by promoting the associative domain, the social domain, the highly social domain, and the cooperative domain. Wong-Powell believes that a skilfully managed process of using digital games helps develop skills essential to the 21st century – character, citizenship, collaboration, communication, creativity, and critical thinking.

The possibilities of using digital materials in small Italian schools were analysed by Giuseppina Rita Jose Mangione and Samuele Calzone. Their chapter includes results from a number of projects in which small Italian schools looked for ways to improve the learning process to reduce the various risks that marginality and isolation bring. The authors prioritized the use of an integrated survey approach, and the projects were investigated through an inductive approach underlying interpretative processes of qualitative analysis, taking into account the materiality dimensions. They concluded that technology becomes a material element that helps to review spatial and traditional objects (contents, activities), making them connected and inter-individual, and strengthens them in terms of the explorations of scientific disciplines.

Marilene Santana dos Santos Garcia and Luciano Frontino de Medeiros, in turn, provide ideas about what should be done in defining the skills and competencies needed for educational curators in the context of TEL. They believe that to use technology in the learning process, it is necessary to know how to use different technologies and to know the content to be learned.

Overall, the book's collection of ideas, research results, and conclusions creates the basis for further discussions on the use of technology options, and requires us to rethink our existing experiences and maybe also help to overcome some prejudices about technology's place and role in the learning process, the digital play and gaming process, and the process of the development of new teaching aids.

In light of the development of digital solutions and changes in information retrieval habits, it is also important to develop new pedagogical solutions, offer new ways of learning, and produce new learning content that can be found in the digital environment.

Appendix A

Vignette- The use of a smartphone in children's role play

A group of three year olds are role playing in the home corner. They are seated on small chairs around a table. The children are pretending to enjoy a meal together. They are playing adjacent to each other and they are playing in silence.

A few moments later, one child suddenly jumps up and says 'Shhh, stop!' She reaches into her pocket and pretends to pull out a smartphone. Swiping her palm a few times, she cups her hand over her ear and begins to speak. 'Hello. Yes, yes, yes. I eating.'

The other children sitting around the table look up from their plates. What was once a silent meal between the children, has now become a more dynamic interaction. One child asks 'Mommy?' Another child asks 'It grandma?'

The child who is now engaged in a full conversation on the smartphone, holds up her pointer finger to signal for silence amongst the group. However, the other children continue to ask their questions. Finally, the child turns her palm outwards to face the group. The children all lean in as they take a closer look at her palm. One child then points and says, 'Oh, it is our daddy.'

The dialogue continues as the children take turns looking at the imagined screen of the smartphone. As the children exchange details of what they are eating and what they plan to do after their meal, some children even attempt to swipe the pretend screen. At which point, the girl turns to the group and says "Finish."

(Chapter 6- Unpacking Digital Play in Children's Play, Jennifer Wong-Powell, p. 89).

Index